AF595016

New Insights on Biofilm Antimicrobial Strategies, 2nd Volume

New Insights on Biofilm Antimicrobial Strategies, 2nd Volume

Editors

Luís Melo
Andreia S. Azevedo

MDPI • Basel • Beijing • Wuhan • Barcelona • Belgrade • Manchester • Tokyo • Cluj • Tianjin

Editors
Luís Melo
CEB - Centre of Biological Engineering
University of Minho
Braga
Portugal

Andreia S. Azevedo
LEPABE - Laboratory for Process Engineering, Environment, Biotechnology and Energy
Faculty of Engineering, University of Porto
Porto
Portugal

Editorial Office
MDPI
St. Alban-Anlage 66
4052 Basel, Switzerland

This is a reprint of articles from the Special Issue published online in the open access journal *Antibiotics* (ISSN 2079-6382) (available at: www.mdpi.com/journal/antibiotics/special_issues/biofilm_2nd).

For citation purposes, cite each article independently as indicated on the article page online and as indicated below:

LastName, A.A.; LastName, B.B.; LastName, C.C. Article Title. *Journal Name* **Year**, *Volume Number*, Page Range.

ISBN 978-3-0365-4802-9 (Hbk)
ISBN 978-3-0365-4801-2 (PDF)

Contents

About the Editors

Luís Melo

Luís D. R. Melo (PhD). With a degree and MSc in Biology and a post-graduation in Methods of DNA analysis, he obtained his PhD in Biomedical Engineering at the University of Minho, where he is currently an Assistant Researcher. Luis DR Melo established his research on exploring phages interaction with bacterial biofilms and has demonstrated the value of phage proteins on the control and detection of pathogenic bacteria. Experience on flow cytometry, genomics, transcriptomics and proteomics led to an increased knowledge on phage/host interactions. He has been involved in different scientific funded projects, being the PI of two. He is also involved on the organization of international practical courses and conferences within this field of knowledge. The quality and innovative character of the research developed, let to the publication of several international refereed papers and book chapters and enabled the creation of a patent.

Andreia S. Azevedo

Andreia S. Azevedo completed her PhD degree, in June 2016, at Chemical and Biological Engineering, from Faculty of Engineering (FEUP), University of Porto, in collaboration with the Nucleic Acid Center, Department of Physics, Chemistry and Pharmacy, University of Southern Denmark (NAC, SDU). Currently, she is a junior researcher at the Laboratory for Process Engineering, Environment, Biotechnology, and Energy (LEPABE/FEUP). Her main research interests include exploring and studying the hybridization properties of nucleic acid mimics (NAMs); and combining NAM-fluorescence in situ hybridization (NAM-FISH) and spectral imaging for multiplex detection/location of microorganisms in multispecies biofilms. She has been involved in different scientifically funded projects, being the PI of one (project reference: POCI-01-0145-FEDER-016678). Andreia S. Azevedo also has a long track-record in supervising/co-supervising MSc and PhD students. She has participated in several scientific dissemination events (seminars, workshops, congresses, courses); including, the organization of international practical courses and conferences. Her work has led to the publication of several papers in peer-reviewed international journals and book chapters.

Editorial

New Insights on Biofilm Antimicrobial Strategies, 2nd Volume

Andreia S. Azevedo [1,2,3,4,*] and Luís D. R. Melo [5,6,*]

1 LEPABE-Laboratory for Process Engineering, Environment, Biotechnology and Energy, Faculty of Engineering, University of Porto, 4200-465 Porto, Portugal
2 ALiCE-Associate Laboratory in Chemical Engineering, Faculty of Engineering, University of Porto, 4200-465 Porto, Portugal
3 i3S-Instituto de Investigação e Inovação em Saúde, University of Porto, 4200-135 Porto, Portugal
4 IPATIMUP-Instituto de Patologia e Imunologia Molecular, Universidade do Porto, 4200-135 Porto, Portugal
5 Centre of Biological Engineering (CEB), University of Minho, 4710-057 Braga, Portugal
6 LABBELS—Associate Laboratory, Braga/Guimarães, Portugal
* Correspondence: asazevedo@fe.up.pt (A.S.A.); lmelo@deb.uminho.pt (L.D.R.M.)

In biofilms, microorganisms are able to communicate together and assemble by themselves, creating a consortium with different properties from the original free-floating microorganisms. In fact, biofilm cells bind strongly to a living or non-living surface, enclosed in a self-produced extracellular matrix that is composed of extracellular polymeric substances. One benefit of this lifestyle is the increased resistance or tolerance to antimicrobial agents (e.g., antibiotics). Hence, research on the development of alternative strategies to prevent and control biofilms is highly relevant for society in terms of human health, industry and the environment. Different approaches to prevent or control biofilms using antibiotic alternative strategies were submitted to this Special Issue.

An important topic is the study of quorum sensing (QS) mechanisms during biofilm development. Daza et al. evaluated the use of alkylglycerols as QS inhibitors. The effect on the biofilm formation of fifteen natural enantiopure alkylglycerols was evaluated on *Chromobacterium violaceum*. The authors observed a dose-dependent response using each alkylglycerol at subinhibitory concentrations, with a maximum response of 97.2% reduction. (2S)-3-O-(cis-13′-docosenyl)-1,2-propanediol was the best QS inhibitor with just 20 μM [1].

The use of chemical compounds, bacteriophages and enzymes are also alternative approaches with the potential to control the bacterial growth and the biofilm development. For instance, Alfarrayeh et al. evaluated the effect of Caffeic Acid Phenethyl Ester (CAPE) on different *Candida* species. Both caspase-dependent and caspase-independent apoptosis were observed after CAPE treatment. The minimum biofilm inhibitory concentration oscillated between 50 and 100 μg/mL. Regarding biofilm control, the authors observed a low to high capability to control mature *Candida* biofilms [2].

In another study, Pereira et al. used gapmers and steric blockers as antisense oligonucleotides against *Escherichia coli*. These molecules were conjugated with vitamin B12 by copper-free azide-alkyne click-chemistry. The authors observed that despite the strong interaction with *E. coli*, conjugates were mostly located on the outer membrane. Only 6–9% reached the cytosol, which was not sufficient to inhibit bacterial growth. The authors concluded that this low internalization of conjugates was affected by *E. coli*'s low uptake for vitamin B12 and further studies are necessary before applying it to infectious biofilms [3].

Additionally, using stainless steel surfaces, González-Gómez et al. studied the efficacy of novel bacteriophages against *E. coli* biofilms. Three bacteriophages isolated from ground beef and poultry liver samples with a podovirus-like morphology were used alone or in combination to treat biofilms for 2, 14 and 48 h. Some very significant bacterial cell count reductions were observed after treatment, with it being important to note that the treatment success was influenced by the bacterial strain used, the used phage, phage concentration and biofilm formation stage [4].

Citation: Azevedo, A.S.; Melo, L.D.R. New Insights on Biofilm Antimicrobial Strategies, 2nd Volume. *Antibiotics* **2022**, *11*, 908. https://doi.org/10.3390/antibiotics11070908

Received: 23 June 2022
Accepted: 4 July 2022
Published: 7 July 2022

The effects of enzymatic and chemical treatments were compared to remove mixed-species biofilms on surfaces simulating the food processing environment. Iñiguez-Moreno et al. used a novel chemical product named Sanicip Bio Control or peracetic acid as chemical agents and protease or α-amylase for enzymatic treatments. Biofilms were formed on stainless steel or polypropylene B and different media were used (whole milk, TSB with meat extract and TSB with chicken egg yolk). Despite the success of some treatments, in general the results were strongly affected by food contact surfaces and the surrounding media [5].

A combined treatment is another interesting strategy; Aljaafari et al. investigated the effect of various thermal shocks treatments (different temperature and times), the effect of different ciprofloxacin hydrochloride concentrations and the interaction of antibiotics and thermal shock on *Pseudomonas aeruginosa* biofilms. In addition, to assess the viability after the thermal shock, the biofilms were subsequently re-incubated under the initial conditions. To generate biofilms with different population densities, structures, and maturities, *P. aeruginosa* biofilms were grown on 4-well dishes at 160 rpm in an incubator (ST biofilms) and in a Drip Flow Reactor (DFR biofilms). The authors concluded that the use of ciprofloxacin does not appear to enhance thermal shock directly. However, they suggested that thermal shock and antibiotics act with strictly orthogonal mechanisms on *P. aeruginosa* biofilms, decreasing the intensity of thermal shock needed [6].

Barros et al. tested the antimicrobial activity of biocide benzalkonium chloride (BAC) immobilized in millimetric aluminum oxide particles, against *Escherichia coli* bacteria in the planktonic state. The alumina particles were functionalized by using a nitrogen precursor (Dopamine (DA)). At the highest particle concentration (3000 mg/L), an inactivation of bacterial cells within 5 min was observed. In addition, a total loss of membrane integrity was achieved after 15 min for all tested concentrations. However, when reusing the Al_2O_3-DA-BAC particles, a higher contact time was needed to reach total inactivation. The authors concluded that Al_2O_3-DA-BAC particles are a suitable antimicrobial agent for the treatment of continuous water systems with minimal environmental and health impacts [7].

Surface engineering approaches are another successful strategy to control biofilms. Sunthar et al. analyzed the antibacterial properties of Cellulose Acetate (CA) reinforced with different weight percentages of aluminum nitride (AlN) composites against *Staphylococcus epidermidis* and *E. coli*. The results showed an effective antibacterial effect when AIN was added in weight percentages >10 wt.%, suggesting the potential application of CA/AlN composites as alternative materials for plastic packaging in the food industry. However, the authors suggest that the degradability and stability of this composite material should be studied in future work [8].

Using a different and innovative approach, Carvalho et al. evaluated the effect of two probiotic strains (*Lactobacillus plantarum* and *Lactobacillus rhamnosus*) in displacing *E. coli* and Staphylococcus aureus pre-formed biofilms. The biofilms were grown under conditions that mimic the urological devices, including silicone surfaces, artificial urine medium and a shear stress similar to those found inside of urinary catheters. This is the first study that demonstrates the ability of *L. plantarum* and *L. rhamnosus* to displace pre-established biofilms of *E. coli* and *S. aureus*. These results showed the potential of these Lactobacillus strains to control the development of biofilms on urinary tract devices [9].

Finally, Esteves et al. provided a review to evaluate the significance of the incorporation of antibacterial coatings in the reduction in the occurrence of bacterial infections and evaluate its effect on dental implant success rate. The authors summarize the diverse strategies proposed to enhance the antibacterial properties of titanium dental implants. This is significant because antibacterial coatings are a promising solution to control and prevent bacterial infections that compromise dental and orthopedic implant success [10].

Author Contributions: A.S.A. and L.D.R.M. wrote and reviewed the manuscript. All authors have read and agreed to the published version of the manuscript.

Acknowledgments: This work was financially supported by: LA/P/0045/2020 (ALiCE), UIDB/00511/2020 and UIDP/00511/2020 (LEPABE), funded by national funds through FCT/MCTES (PIDDAC); Project POCI-01-0145-FEDER-030431 (CLASInVivo), funded by FEDER funds through COMPETE2020—Programa Operacional Competitividade e Internacionalização (POCI) and by national funds (PIDDAC) through FCT/MCTES. This study was supported by the Portuguese Foundation for Science and Technology (FCT) under the scope of the strategic funding of the UIDB/04469/2020 unit and the project DeathTrigger PTDC/BIA-MIC/2312/2020.

Conflicts of Interest: The authors declare no conflict of interest.

References

1. Chaverra Daza, K.E.; Silva Gómez, E.; Moreno Murillo, B.D.; Mayorga Wandurraga, H. Natural and Enantiopure Alkylglyc-erols as Antibiofilms Against Clinical Bacterial Isolates and Quorum Sensing Inhibitors of Chromobacterium violaceum ATCC 12472. *Antibiotics* **2021**, *10*, 430. [CrossRef] [PubMed]
2. Alfarrayeh, I.; Pollák, E.; Czéh, Á.; Vida, A.; Das, S.; Papp, G. Antifungal and Anti-Biofilm Effects of Caffeic Acid Phenethyl Ester on Different *Candida* Species. *Antibiotics* **2021**, *10*, 1359. [CrossRef] [PubMed]
3. Pereira, S.; Yao, R.; Gomes, M.; Jørgensen, P.; Wengel, J.; Azevedo, N.; Santos, R.S. Can Vitamin B12 Assist the Internalization of Antisense LNA Oligonucleotides into Bacteria? *Antibiotics* **2021**, *10*, 379. [CrossRef] [PubMed]
4. González-Gómez, J.P.; González-Torres, B.; Guerrero-Medina, P.J.; López-Cuevas, O.; Chaidez, C.; Avila-Novoa, M.G.; Gutiérrez-Lomelí, M. Efficacy of Novel Bacteriophages against *Escherichia coli* Biofilms on Stainless Steel. *Antibiotics* **2021**, *10*, 1150. [CrossRef] [PubMed]
5. Iñiguez-Moreno, M.; Gutiérrez-Lomelí, M.; Avila-Novoa, M. Removal of Mixed-Species Biofilms Developed on Food Contact Surfaces with a Mixture of Enzymes and Chemical Agents. *Antibiotics* **2021**, *10*, 931. [CrossRef] [PubMed]
6. Aljaafari, H.; Gu, Y.; Chicchelly, H.; Nuxoll, E. Thermal Shock and Ciprofloxacin Act Orthogonally on *Pseudomonas aeruginosa* Biofilms. *Antibiotics* **2021**, *10*, 1017. [CrossRef] [PubMed]
7. Barros, A.; Pereira, A.; Melo, L.; Sousa, J. New Functionalized Macroparticles for Environmentally Sustainable Biofilm Control in Water Systems. *Antibiotics* **2021**, *10*, 399. [CrossRef] [PubMed]
8. Sunthar, T.P.M.; Boschetto, F.; Doan, H.N.; Honma, T.; Kinashi, K.; Adachi, T.; Marin, E.; Zhu, W.; Pezzotti, G. Antibacterial Property of Cellulose Acetate Composite Materials Reinforced with Aluminum Nitride. *Antibiotics* **2021**, *10*, 1292. [CrossRef] [PubMed]
9. Carvalho, F.M.; Mergulhão, F.J.M.; Gomes, L.C. Using Lactobacilli to Fight *Escherichia coli* and *Staphylococcus aureus* Biofilms on Urinary Tract Devices. *Antibiotics* **2021**, *10*, 1525. [CrossRef] [PubMed]
10. Esteves, G.M.; Esteves, J.; Resende, M.; Mendes, L.; Azevedo, A.S. Antimicrobial and Antibiofilm Coating of Dental Implants—Past and New Perspectives. *Antibiotics* **2022**, *11*, 235. [CrossRef] [PubMed]

MDPI

Article

Natural and Enantiopure Alkylglycerols as Antibiofilms Against Clinical Bacterial Isolates and Quorum Sensing Inhibitors of *Chromobacterium violaceum* ATCC 12472

Klauss E. Chaverra Daza [1,2], Edelberto Silva Gómez [2], Bárbara D. Moreno Murillo [3] and Humberto Mayorga Wandurraga [3,*]

1. Posgrado Interfacultades de Microbiología, Facultad de Ciencias, Universidad Nacional de Colombia, Av. Carrera 30 # 45-03, Edif. 224, Bogotá 11011, Colombia; kchaverra@unal.edu.co
2. Grupo de Productos Naturales Vegetales Bioactivos y Química Ecológica, Laboratorio de Asesorías e Investigaciones en Microbiología, Departamento de Farmacia, Facultad de Ciencias, Universidad Nacional de Colombia, Av. Carrera 30 # 45-03, Edif. 450, Bogotá 11011, Colombia; esilvag@unal.edu.co
3. Grupo de Productos Naturales Vegetales Bioactivos y Química Ecológica, Departamento de Química, Facultad de Ciencias, Universidad Nacional de Colombia, Av. Carrera 30 # 45-03, Edif. 451, Bogotá 11011, Colombia; bdmorenom@unal.edu.co

* Correspondence: hmayorgaw@unal.edu.co; Tel.: +57-1-316-5000 (ext. 14440)

Abstract: Resistance mechanisms occur in almost all clinical bacterial isolates and represent one of the most worrisome health problems worldwide. Bacteria can form biofilms and communicate through quorum sensing (QS), which allow them to develop resistance against conventional antibiotics. Thus, new therapeutic candidates are sought. We focus on alkylglycerols (AKGs) because of their recently discovered quorum sensing inhibition (QSI) ability and antibiofilm potential. Fifteen natural enantiopure AKGs were tested to determine their effect on the biofilm formation of other clinical bacterial isolates, two reference strains and their QSI was determined using *Chromobacterium violaceum* ATCC 12472. The highest biofilm inhibition rates (%) and minimum QS inhibitory concentration were determined by a microtiter plate assay and ciprofloxacin was used as the standard antibiotic. At subinhibitory concentrations, each AKG reduced biofilm formation in a concentration-dependent manner against seven bacterial isolates, with values up to 97.2%. Each AKG displayed QSI at different levels of ability without affecting the growth of *C. violaceum*. AKG (2*S*)-3-*O*-(*cis*-13′-docosenyl)-1,2-propanediol was the best QS inhibitor (20 μM), while (2*S*)-3-*O*-(*cis*-9′-hexadecenyl)-1,2-propanediol was the least effective (795 μM). The results showed for the first time the QSI activity of this natural AKG series and suggest that AKGs could be promising candidates for further studies on preventing antimicrobial resistance.

Keywords: antimicrobial resistance; natural alkylglycerols; ether lipids; 1-*O*-alkyl-sn-glycerols; antibiofilm activity; quorum sensing inhibition

Citation: Chaverra Daza, K.E.; Silva Gómez, E.; Moreno Murillo, B.D.; Mayorga Wandurraga, H. Natural and Enantiopure Alkylglycerols as Antibiofilms Against Clinical Bacterial Isolates and Quorum Sensing Inhibitors of *Chromobacterium violaceum* ATCC 12472. *Antibiotics* **2021**, *10*, 430. https://doi.org/10.3390/antibiotics10040430

Academic Editors: Luís Melo and Andreia Azevedo

Received: 10 February 2021
Accepted: 2 April 2021
Published: 13 April 2021

Publisher's Note: MDPI stays neutral with regard to jurisdictional claims in published maps and institutional affiliations.

1. Introduction

Antibiotics were discovered in the early 20th century and have been widely used in the treatment of bacterial infections. However, the indiscriminate and excessive usage of antibiotics has led to the development of antimicrobial resistance, which has steadily decreased the effectiveness of current antibacterial therapies [1,2]. Resistance mechanisms occur in almost all clinical isolates of bacteria and recurrent infections caused by persistent bacteria hamper the successful treatment of infections [3]. This situation highlights the urgency of identifying new therapeutic candidates and less toxic treatment targets. Consequently, considerable research has focused on developing novel strategies to control bacterial diseases [4]. Infectious diseases cause the death of more than 16 million people annually and at least 65% of these cases are linked to bacterial communities that proliferate by forming biofilms [5,6]. Biofilms consist of microorganisms that are attached to a substratum and embedded in a matrix of extracellular polymers that protects the cells from harsh environmental conditions, including disinfectant treatments and antibiotics [7]. Bacterial biofilms are developed by a cell communication process called quorum sensing (QS), through which Gram-negative and Gram-positive bacteria synchronize their behaviors in a cell density-dependent mode by a series of signal molecules termed autoinducers, thereby mediating the production and secretion of virulence traits. In general, QS is a regulatory mechanism that promotes the establishment of infection, expression of virulence factors, formation of biofilm and development of resistance [8,9]. Biofilms are involved in many clinical infections and during biofilm infection, simultaneous activation of both innate and acquired host immune responses may occur; however, these responses do not eliminate the biofilm pathogen but rather accelerate collateral tissue damage [10]. A strategy for controlling microbial infections involves the use of agents that inhibit bacterial pathogenicity (e.g., production of virulence factors and biofilm formation) rather than targeting growth-dependent mechanisms that will inevitably lead to the development of microbial resistance [11,12]. QS inhibitors do not kill bacteria or inhibit bacterial growth; rather, they quench QS-regulated pathogenic behaviors, which inhibits bacterial pathogenesis and does not easily induce antibiotic-resistant mutations [13,14]. The communication of individual cells is essential for the formation of biofilms; therefore, blocking this QS process is an important goal for the control of biofilm infections and represents a new method of combating antimicrobial resistance [15].

Efforts have been made to disrupt biofilms by inhibiting the QS system and many natural and synthetic molecules exhibiting quorum sensing inhibition (QSI) have been identified with potential therapeutic approaches [2]. However, natural products continue to be a prolific source of drugs [13,16] and the rich biodiversity in the marine environment suggests that it represents an enormous resource of novel therapeutic candidates with antibiofilm and QSI activities [17,18]. Marine sources, such as shark liver oil, contain high levels of alkylglycerols (AKGs), which are minor constituents of bacteria, protozoa, fungi, higher plants, animals and humans [19,20]. Natural AKGs called 1-*O*-alkyl-sn-glycerols are structurally glycerol ether lipids that occur as a mixture of alkyl chains of varying lengths and unsaturation levels and as pure enantiomers with an *S* configuration at the asymmetric carbon [21,22]. AKGs generally occur in their diacylated form as 1-*O*-alkyl-2,3-diacyl-sn-glycerols and they are also natural precursors of their derivatives ether phospholipids, which participate in the structure and function of cell membranes [20,23]. AKGs display a broad range of beneficial effects on human health and can be used to treat gastric ulcers and colon inflammation; stimulate hematopoiesis, immunological defense and vaccination efficiency; reduce tumor growth, metastasis, radiotherapy side effects, obesity and oxidative stress; increase therapeutic molecule permeability through the hematoencephalic barrier and sperm motility; and regulate cell differentiation and neuropathic pain [19,20,23–28]. Their activity spectrum also includes antifungal [29] and antibacterial properties [30–35].

In preliminary works to identify bioactive compounds from marine invertebrates [36], extracts from two soft coral *Eunicea* species and an AKG purified from them, which is known as batyl alcohol **8**, showed antibiofilm capacity against bacteria isolated from marine fouled surfaces [37,38]. Two more AKGs demonstrated growth inhibition of marine biofilm-forming bacteria, which are regarded as potential antifouling compounds [39,40]. In addition, a mixture of AKGs from *Eunicea* together with the AKG (2*S*)-3-*O*-dodecyl-1,2-propanediol **4** and (2*S*)-3-*O*-tetradecyl-1,2-propanediol **5** also showed QSI activity in *C. violaceum* ATCC 31532 [41]. Fifteen natural enantiopure AKGs were synthesized and they depicted promising ability against two biofilm-forming bacteria, clinical isolate *Staphylococcus aureus* 91 and *Pseudomonas aeruginosa* ATCC 15442 [42]. Thus, in the current study, we selected this natural AKG series to evaluate their antibiofilm activity against other clinical bacterial isolates, other reference strains and assess their QSI ability using *C. violaceum* ATCC 12472.

2. Results

2.1. Structures of Natural and Enantiopure Alkylglycerols

The AKGs included in this study, **1–6**, **8**, **12** and **14**, are of saturated aliphatic chains, while AKGs **7**, **9**, **13** and **15** are mono-unsaturated, **10** is di-unsaturated and **11** is a tri-unsaturated AKG. They have a chain from 6 to 22 carbons atoms and between 176 to 400 amu (Figure 1).

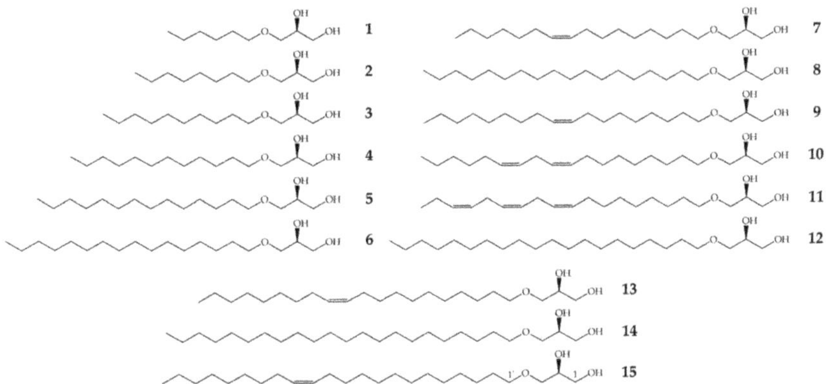

Figure 1. Structures of alkylglycerols used in this study.

2.2. Minimal Inhibitory Concentration (MIC) of Alkylglycerols

An initial in vitro screening was performed to evaluate the MIC of all AKGs (Figure 1) against the bacterial strains included in this study and the results are summarized in Table 1.

Table 1. Minimal inhibitory concentration values (µM) of the examined alkylglycerols.

Strains	Alkylglycerols [a]									
	CPX [b]	2	3	4	5	6	7	9	10	11
K. pneumoniae 792	0.09	>1224	>1076	>960	>867	>790	>795	>730	>734	>738
E. cloacae 250	48.2	>1224	>1076	>960	>867	395	>795	>730	>734	>738
E. coli 667	0.09	>1224	>1076	>960	>867	>790	>795	>730	>734	>738
P. aeruginosa 740	1.51	>1224	>1076	>960	>867	>790	>795	>730	>734	>738
P. mirabilis 26	0.09	>1224	1076	>960	>867	>790	>795	>730	>734	>738
C. violaceum ATCC 12472	0.18	>1224	269	>960	>867	>790	>795	730	>734	>738
E. faecalis 12	0.51	>1224	538	119	433	>790	199	365	734	>738
E. gallinarum 662	48.2	>1224	>1076	119	>867	>790	>795	>730	>734	>738
S. epidermidis ATCC 12228	0.39	>1224	538	119	108	>790	795	>730	734	>738
S. aureus ATCC 6538	0.18	612	67	30	27	>790	99	>730	367	369

[a] Alkylglycerols **1**, **8**, **12**, **13**, **14** and **15** had a MIC values >1418, >726, >671, >675, >624 and >627 µM respectively for all tested bacterial strains. All experiments were performed with a maximum of 0.9% (*v*/*v*) DMSO in MHB medium and it did not interfere with bacterial growth. [b] CPX = ciprofloxacin. CPX was the antibacterial agent used as the standard. Values are the means of triplicate determinations.

2.3. Alkylglycerols Ability to Inhibit of Biofilm Formation

The clinical bacterial isolates and the reference strains were selected because of their biofilm-forming ability evidenced in this study. All examined AKGs were able to inhibit biofilm formation of most clinical isolates and reference strains and showed different capacities against each microorganism assayed. However, the inhibition intensity varied depending on the compound concentration (Figure 2a,b). The highest values of the biofilm inhibitory percentage were observed at AKGs concentrations corresponding to 0.50 MIC. Remarkable biofilm inhibition ratios of several AKGs above 50% were found for *E. cloacae* 250, *P. aeruginosa* 740 and *S. aureus* ATCC 6538, while a low response below 27% was found for *S. epidermidis* ATCC 12228. *K. pneumoniae* 792 and *S. aureus* ATCC 6538 were the only strains active against all AKGs.

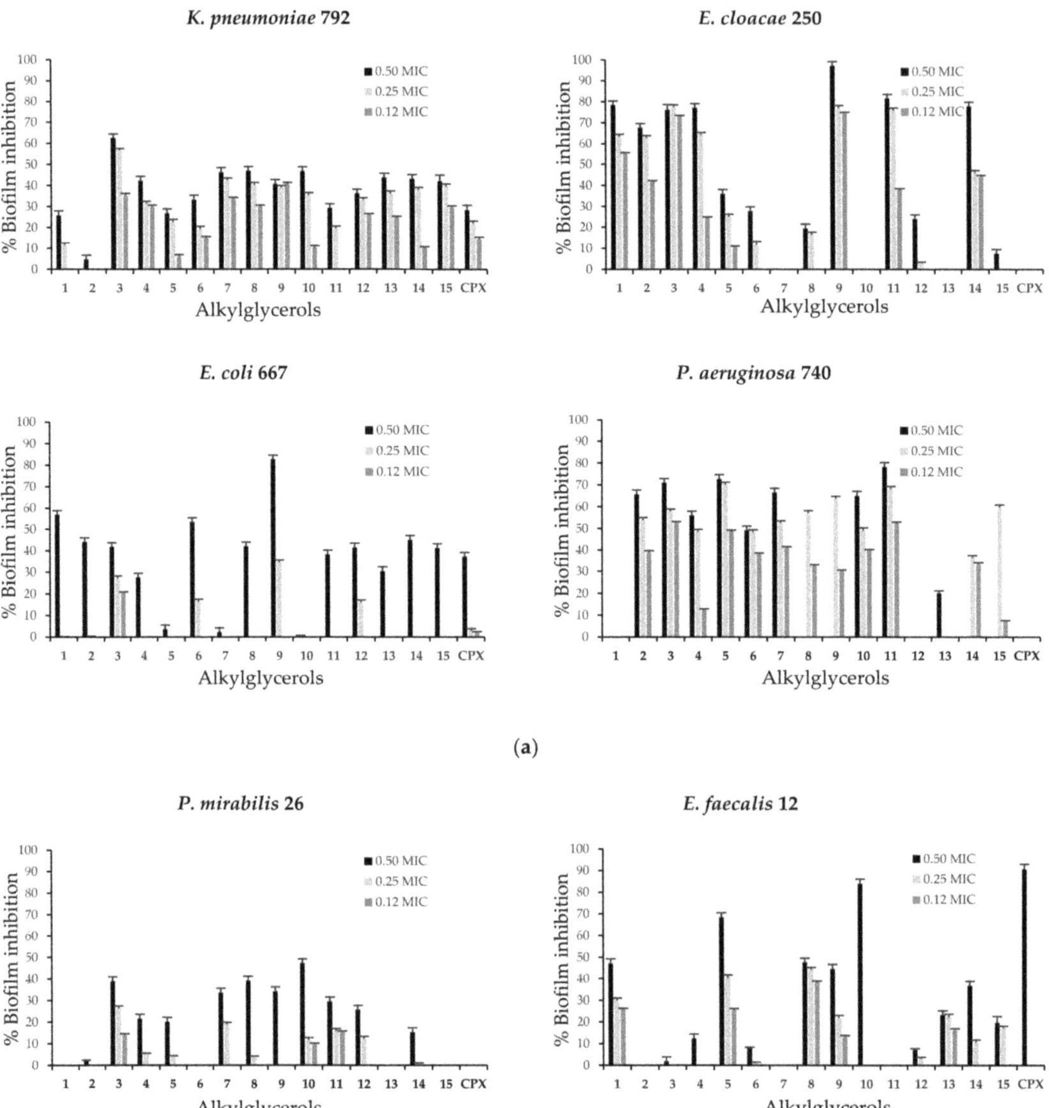

Figure 2. *Cont.*

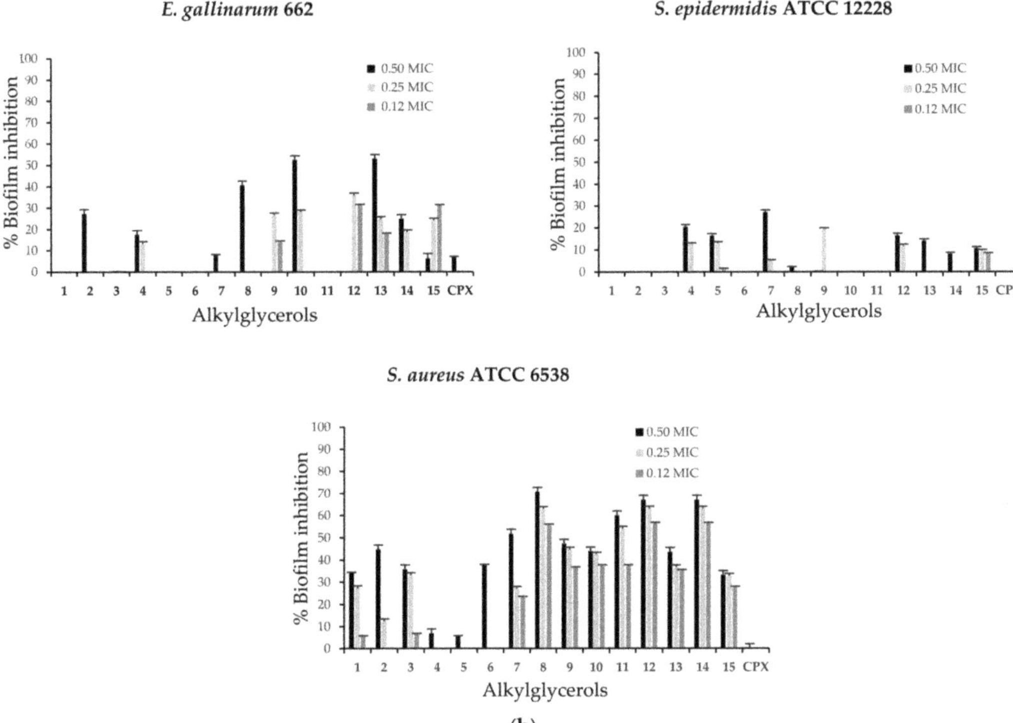

Figure 2. (**a**) Biofilm inhibition represented as percentages of enantiopure alkylglycerols against clinical bacterial isolates: *K. pneumoniae* 792, *E. cloacae* 250, *E. coli* 667 and *P. aeruginosa* 740 at different subinhibitory concentrations. (**b**) Biofilm inhibition represented as percentages of enantiopure alkylglycerols against clinical bacterial isolates: *P. mirabilis* 26, *E. faecalis* 12, *E. gallinarum* 662 and reference strains *S. epidermidis* ATCC 12228 and *S. aureus* ATCC 6538 at different subinhibitory concentrations. CPX was subject to the same assays. The results are expressed as the mean of three independent experiments and error bars represent the standard deviation. Biofilm formation was significantly reduced compared to untreated control biofilms of each of these bacteria at the tested concentrations of the compounds.

AKG **9** showed a reduction in biofilm formation of all strains exhibiting the best antibiofilm activity and the highest inhibition ratios were observed for the Gram-negative clinical isolates, *E. cloacae* 250 (97.2%), *E. coli* 667 (82.6%) at 0.50 MIC and *P. aeruginosa* 740 (63.7%) at 0.25 MIC. AKG **11** showed great biofilm reduction for *E. cloacae* 250 (81.6%) and *P. aeruginosa* 740 (78.2%) at 0.50 MIC. The other AKGs were less effective.

K. pneumoniae 792 was inhibited by AKG **3** up to 62.5% and AKGs **7**, **8** and **10** by close to 46.5% at 0.50 MIC. In *E. cloacae* 250 in addition to AKGs **9** and **11**, AKGs **1**, **3**, **4** and **14** showed effectiveness and reached inhibition rates of approximately 76% at 0.50 MIC. Other compounds, such as **1** and **6**, decreased biofilm formation by *E. coli* 667 by up to 56.9% and AKGs **2**, **3**, **8**, **12**, **14** and **15** reached almost 45% at 0.50 MIC.

In *P. aeruginosa* 740, biofilm inhibition was also shown by AKGs **3** and **5** at up to 72.6% at 0.50 MIC and a lower response was observed with other compounds. At 0.50 MIC compounds **10** (47.3%), **3** and **8** were the most active against *P. mirabilis* 26. In the case of Gram-positive strains, *E. faecalis* 12 was most susceptible to AKGs **10**, **5** and **1** and an inhibition rate of 83.9% was found for **10**. In addition, for *E. gallinarum* 662 moderate inhibition by AKGs **13** (52.9%), **10** (52.4%) and **8** (40.7%) was observed at 0.50 MIC. For the reference strain *S. epidermidis* ATCC 12228, compounds **7** (27.0%) and **4** (20.3%) at 0.50 MIC exhibited low activity, whereas for *S. aureus* ATCC 6538 showed inhibition up to 70.6% for

AKG **8** at 0.05 MIC and up to 66.9% for compounds **12** and **14**. In comparison, CPX, which is the antibacterial agent and subjected to same treatment, only inhibited biofilm formation of *E. faecalis* 12 up to 90.7%, *E. coli* 667 up to 37.2% and *K. pneumoniae* 792 up to 28.2% at 0.50 MIC.

2.4. Quorum Sensing Inhibition Activity of Natural Alkylglycerols

To determine the effect on the QSI activity using *C. violaceum* ATCC 12472 without interference from the antibacterial activity of the tested compounds, the AKGs were analyzed at subinhibitory concentrations. All AKGs were capable of inhibiting violacein production with varying degrees of action depending on the aliphatic chains on their structures (Table 2). Notably, the highest QSI activity was exhibited by AKG **15**, with a minimum concentration of 20 μM. To a lesser extent, compounds **3**, **4**, **5** and **10** showed relevant activity and reached a reduction in QS up to an inhibitory concentration of 135, 120, 109 and 92 μM respectivelly, followed by AKG **6** at a minimum concentration of 197 μM. In regard to the majority of AKGs, the potential of QSI was lower for **1**, **2**, **8**, **9**, **11**, **12**, **13** and **14** at a concentration varied from 312 to 709 μM, while *C. violaceum* ATCC 12472 was less sensitive for the inhibition of QS to AKG **7** (795 μM).

Table 2. Minimum quorum sensing inhibitory concentration (μM) of alkylglycerols against *C. violaceum* ATCC 12472.

Values. μM [a]	Alkylglycerols														
	1	**2**	**3**	**4**	**5**	**6**	**7**	**8**	**9**	**10**	**11**	**12**	**13**	**14**	**15**
	709	611	135	120	109	197	795	363	365	92	369	335	337	312	20

[a] Minimum quantity μM of compound required to inhibit violacein pigment. Values represent the means from three independent experiments.

The bacterial viability of *C. violaceum* ATCC 12472 was also assessed for both the positive control and AKG-treated samples and the results in terms of colony-forming units/mL (CFU/mL) did not show significant differences between the groups. Therefore, the reduction in violacein production was attributed to the effect of the AKGs.

3. Discussion

Bacteria have developed multiple mechanisms to adapt to changes in the environment, especially the widespread use of antibiotics and these mechanics allow them to avoid adverse conditions and maintain their pathogenesis. One of the mechanisms is QS, a cell-to-cell communication method that depends on the cell density to control collective behavior [1,9]. With the regulation of QS, bacteria can form virulence factors such as toxins, motility, enzymes, proteins and biofilm development to survive in disadvantageous circumstances [2,8]. Biofilms are another critical reason for microbial resistance and they consist of bacteria embedded in a self-produced extracellular matrix of polysaccharides, fibrins, lipids, proteins and DNA that can reduce the effect of antibiotics [7,10,43]. Consequently, there is an urgent search for new candidates and less toxic treatments to control pathogenic resistant bacteria [3,4].

To identify the inhibitors of biofilm formation from bioactive extracts of two Caribbean soft corals of *Eunicea* genus, we isolated an AKG from these species called batyl alcohol **8**, which was the most active inhibitor [37,38]. This class of natural AKGs that had not been previously studied for antibiofilm activity was selected and great variety of their compounds were evaluated. AKGs are bioactive compounds found in low concentrations in marine sources, such as fishes, sponges and corals and at trace levels in plants and mammal, including humans [19,20,23–28]. In nature, they are enantiomerically pure constituents with configuration *S* at the stereogenic center [21,22], which motivated us to obtain several of their naturally occurring structures by enantiomeric synthesis [41].

In this study, fifteen previously synthesized enantiopure AKGs [42] were tested to determine their effect on the biofilm formation of seven clinical bacterial isolates and two reference strains and evaluate their QSI ability using *C. violaceum* ATCC 12472 at subinhibitory concentrations. These AKGs displayed weak antibacterial activity compared with CPX (MIC from 0.09 to 48.2 μM) against all strains at MIC values in the range of 27 to >1418 μM in a microtiter plate assay (Table 1, Figure 1). In this way, only saturated AKGs with an intermediate length of their aliphatic chain such as compounds **3**, **4** and **5** showed in vitro moderate antibacterial activity toward some Gram-positive strains, of them AKGs **4** and **5** were the most active against *S. aureus* ATCC 6538 (MIC of 30 and 27 μM). With less effectiveness against the most bacterial strains resulted the AKGs **2**, **6**, **7**, **9**, **10** and **11** between 99 and 1224 μM, while the AKGs, **1**, **8**, including **12**, **13**, **14** and **15**, the latter which have 20 or more carbon atoms at the aliphatic chain were much less active against all bacterial strains with a MIC from 624 to >1418 μM. Furthermore, the AKGs showed low bacteriostatic effects on the clinical isolates and reference strains because evaluation of the viability by seeding in fresh culture medium led to the recovery of each bacterial species (data not shown). The above is in accordance with other authors and our preliminary works [41,42]. In a previous study, the authors showed that racemic **4** (*rac*-**4**), known as *rac*-1-*O*-dodecylglycerol, inhibited the growth of *Enterococcus faecium* and *Streptococcus mutans* primarily by stimulating autolysin activity and interfering with cell wall synthesis [30–32]. Racemic-**4** inhibits *S. aureus* and toxic shock syndrome toxin 1 production, although the mechanisms of action have not been characterized [35] and produces growth inhibition of members of two yeast genera, *Candida* and *Cryptococcus* [29]. AKGs **1** and **2** are able to prevent the growth of *Chlamydia trachomatis* [33], while AKG **8** show antimicrobial activity toward some marine bacterial species [34,38].

We found that all AKGs at subinhibitory amounts of 0.50, 0.25 and 0.12 MICs were capable of significantly inhibiting biofilm formation by up to 97.2% in both Gram-negative and Gram-positive clinical isolates and in the reference strains, although they had different efficiencies against each strain. (Figure 2a,b). This compound concentration-dependent antibiofilm activity is also dependent on the bacterial susceptibility of each strain to the respective structure. Consistent with the abovementioned findings, we previously observed that most of these AKGs reduced biofilm formation with different capacities against other two strains and the rates were 23.5–99.8% for the clinical isolate *S. aureus* 91 and 14.1–64.0% for *P. aeruginosa* ATCC 15442 [42]. Our additional results against these nine bacterial species considerably broaden the understanding of the antibiofilm abilities of AKGs.

The QSI activity of the AKGs in this work was explored by employing *C. violaceum* ATCC 12472 as a biological model and subinhibitory concentrations. All enantiopure AKGs significantly inhibited violacein production at minimum QS inhibitory concentrations in the range of 20 to 795 μM (Table 2). Surprisingly, among the tested compounds, the most potent QSI activity was noted for the highest molecular weight and unsaturated AKG, namely, (2*S*)-3-*O*-(*cis*-13′-docosenyl)-1,2-propanediol **15**, with a minimum concentration of 20 μM, while (2*S*)-3-*O*-(*cis*-9′-hexadecenyl)-1,2-propanediol **7** turned out to have the lowest QSI capacity (795 μM). In addition, AKGs **3**, **4**, **5** and **10** reached a reduction in QS at a minimum concentration of 135, 120, 109 and 92 μM, respectively. Neither AKG affected the growth of *C. violaceum* ATCC 12472. A compound that can to interfere with QS without affecting cell growth is considered a promising inhibitor [11,13,14]. The QSI effects reported in the current study support our preliminary results, where the disc diffusion assay in other strain, *C. violaceum* ATCC 31532, showed that AKG **4** exhibited the same minimum inhibition halo at 20 μg/disc as kojic acid, which is a positive control used as known inhibitor of QS systems, moreover, AKG **5** produced a minimum inhibitory halo at 50 μg/disc [41].

Although it was not possible to deduce a common relationship pattern between the structure and antibiofilm activity or between the structure and QSI activity, each strain positively responded with the aliphatic chain nature of the corresponding AKG. Some related natural molecules, such as fatty acids, are antibacterial agents at high concentrations and have great potential to attenuate microbial biofilm formation and virulence at low concentrations by modulating the QS systems without generating drug resistance [44]. Many studies have been performed to discover natural compounds that could combat pathogenic bacteria. Many natural products with antibiofilm activity have been identified from plants, microbes and marine life and they include various compounds classes, such as alkaloids, fatty acids, organosulfurs, cyclic compounds, phenolics, steroids, terpenoids and other aliphatic compounds. These compounds include: elligic acid glycosides, hamamelitannin, carolacton, skyllamycins, promysalin, phenazines, bromoageliferin, flustramine C, meridianin D and brominated furanones [18,45–50]. The most significant groups of plant-derived QS inhibitors belong to terpenes, terpenoids, phenylpropanoids, acid derivatives, diarylheptanoids, coumarins, flavonoids and tannins [46,51,52]. Microorganisms and marine organisms produce a range of molecules that showing remarkable QSI activity. These molecules belong to organic compounds classes: derived alkaloids, terpenes, peptides, polyketides, derived carbohydrates, sesterterpenes, halogenated furanones, cembranoids, halogenated alkaloids and phenolics and they include compounds, such as kojic acid, phenethylamides, aculenes, meleagrin, malyngamides, manoalides, hymenialdisin, betonicines, floridoside and cembrane diterpenoids [18,53,54].

Among the Gram-negative bacteria susceptible to AKGs in the present study, *K. pneumoniae* is a pathogenic bacterium that shows high rates of antibiotic resistance through biofilm formation [55]; *E. cloacae* is an emerging pathogen responsible for various diseases, including respiratory tract infections associated with biofilm formation [56]; *E. coli* pathogenic strains cause problematic biofilm infections and employ a furanosyl borate diester (AI-2)-based QS system to regulate two of its most studied phenotypes, virulence and biofilm formation [57]; another pathogen that has been well studied and is frequently associated with a wide range of severe infections is *P. aeruginosa*, which includes cystic fibrosis pneumonia and chronic obstructive pulmonary disease-related infections. The high level of *P. aeruginosa* pathogenicity is mostly controlled by QS, which regulates genetic expression, with 30% encoding virulence factor production and biofilm formation based on the release and sensing of the signal molecules homoserine lactones [6,8,9,11,57]; *P. mirabilis* is another opportunistic pathogen implicated in various human diseases of the respiratory tract and gastrointestinal tract and such infections are complicated by the unique ability of this bacterium to form crystalline biofilms that protect it from antibiotics and the host immune response [58]; among the most known Gram-negative bacteria, *C. violaceum* is a biosensor strain used for QS research that produces the pigment violacein in response to QS-regulated gene expression and it has been widely used to study the inhibition of QS [4,59]. Among Gram-positive bacteria, *E. faecalis* has become one of the most prevalent multidrug-resistant hospital pathogens and its QS system is closely related to biofilm development and virulence production [60]; *E. gallinarum* is less commonly identified in humans but is responsible for bloodstream, urinary tract and surgical wound infections and is resistant to many antibiotics [61]; the most important virulence factor of *S. epidermidis* is considered to be its ability to form biofilms on implanted biomaterials, these films are difficult to treat because of the increased resistance to both antibiotics and the human immune system [62]; in some cases, *S. aureus* causes a wide range of human infections on skin and soft tissues, as well as life-threatening pneumonia, bacteremia, osteomyelitis and toxic shock syndrome. *S. aureus* uses an autoinducing peptide signal to mediate QS and its pathogenicity depends on its ability to produce biofilms and virulence factors, such as different toxins, that contribute to invasion of the host and bacterial spread, [35,63].

In the current research, the differences in susceptibility among the bacteria tested to each AKG suggested that a large range of these structures with different molecular sizes, unsaturation and polarities have both antibiofilm and QSI activities, which result in much more effectiveness than their respective antibacterial effects. However, the most noteworthy information obtained from the present work revealed for the first time that this variety of natural AKGs are promising QS inhibitors. Since the target of QSI is bacterial virulence factors and bacterial biofilms, not viability, there is less chance that the bacteria will develop resistance to QS inhibitors [5,12,63]. For the above reasons, AKGs could be considered potential candidates to control pathogenic bacterial diseases and thus warrant further experiments involving other model systems to establish their action mechanisms and the extent of their efficacy against antimicrobial resistance.

4. Materials and Methods

4.1. General

4.1.1. Screened Compounds

Fifteen AKGs previously synthesized in our laboratory in enantiomerically pure forms, such as the naturally occurring (Figure 1), were included in this study [42]: (2*S*)-3-*O*-hexyl-1,2-propanediol **1,** (2*S*)-3-*O*-octyl-1,2-propanediol **2**, (2*S*)-3-*O*-decyl-1,2-propanediol **3**, (2*S*)-3-*O*-dodecyl-1,2-propanediol **4**, (2*S*)-3-*O*-tetradecyl-1,2-propanediol **5**, (2*S*)-3-*O*-hexadecyl-1,2-propanediol **6**, (2*S*)-3-*O*-(*cis*-9′-hexadecenyl)-1,2-propanediol **7**, (2*S*)-3-*O*-octadecyl-1,2-propanediol **8**, (2*S*)-3-*O*-(*cis*-9′-octadecenyl)-1,2-propanediol **9**, (2*S*)-3-*O*-(*cis*,*cis*-9′,12′-octadecadienyl)-1,2-propanediol **10**, (2*S*)-3-*O*-(*cis*,*cis*,*cis*-9′,12′,15′-octadecatrienyl)-1,2-propanediol **11**, (2*S*)-3-*O*-(eicosyl)-1,2-propanediol **12**, (2*S*)-3-*O*-(*cis*-11′-eicosenyl)-1,2-propanediol **13**, (2*S*)-3-*O*-(docosyl)-1,2-propanediol **14** and (2*S*)-3-*O*-(*cis*-13′-docosenyl)-1,2-propanediol **15**.

4.1.2. Bacterial Strains and Culture Conditions

The following bacterial strains were used during our antibacterial, antibiofilm and QSI experiments. The Gram-negative clinical isolates included *Klebsiella pneumoniae* 792, *Enterobacter cloacae* 520, *Escherichia coli* 667, *Pseudomonas aeruginosa* 740 and *Proteus mirabilis* 26. The Gram-negative reference strain *Chromobacterium violaceum* 12472 was donated by Prof. Catalina Arevalo from the Universidad Nacional de Colombia. The Gram-positive clinical isolates include *Enterococcus faecalis* 12 and *Enterococcus gallinarum* 662 and reference strains *Staphylococcus epidermidis* ATCC 12228 and *Staphylococcus aureus* ATCC 6538. The clinical strains were previously isolated from hospital patients and provided by the Hospital of Neiva (Huila), Hospital of Tunal, Hospital of Engativá and Universidad del Bosque (Bogotá) [41] and together with the other reference strains, they were obtained from our microbiology laboratory collection [41,42].

The strains were grown for 24 h at 37 °C in yeast extract malt extract dextrose agar (YMD, Merck, Darmstadt, Germany) and the inoculum was prepared in Mueller-Hinton broth (MHB, Merck), supplemented with glycerol (15% *w*/*v*) and kept at −20 °C until use. In addition, *C. violaceum* 12472 was freshly cultured for 24 h at 22 °C in lysogeny broth (LB, Merck) supplemented with kanamycin 100 μg/mL (Merck) before use.

4.2. Determination of the Minimum Inhibitory Concentration of Screened Compounds

The MIC of AKGs was determined in a microbroth dilution assay according to the Clinical and Laboratory Standards Institute guidelines with slight modifications [42]. Briefly, the inoculum (100 μL, 1×10^4 CFU/mL) was mixed with AKG dissolved in 3.6% dimethyl sulfoxide (DMSO, Merck) solution in water to final concentrations of 250, 125, 62.5, 31.25, 15.63, 7.81, 3.91, 1.95, 0.98 or 0.49 μg/mL in MHB, with up to 200 μL added to each well. All experiments were conducted using a maximum of 0.9% (*v*/*v*) DMSO in medium. The plates were incubated at 37 °C for 24 h under aerobic conditions, while *C. violaceum* 12472 was incubated at 22 °C. Ciprofloxacin (CPX, Merck) was used as the standard antibacterial agent (stock concentrations of 64 μg/mL). MIC was defined as the lowest concentration of AKGs to completely inhibit bacterial growth, which was indicated by a lack of visual turbidity. An inoculated well with solvent in culture medium (positive control) and a well containing only medium (negative control) were included. All experiments were performed in triplicate.

4.3. Effectiveness of Alkylglycerols on the Inhibition of Biofilm Formation

The ability of AKGs to prevent biofilm formation of bacterial strains included in this study was evaluated on 96-well polystyrene microplates (Techno Plastic Products, Trasadingen, Switzerland) according to the method described by Stepanović et al. [64], with some modifications [42]. Briefly, 100 µL of the bacterial strain grown in MHB medium supplemented with 1% glucose was inoculated in each well to 10^5 CFU/mL in the presence of 100 µL subinhibitory concentrations of each AKG dissolved in aqueous solution of DMSO, equivalent to 0.50, 0.25 and 0.12 of MIC or of the highest concentration assayed during the MIC evaluation (250 µg/mL). The DMSO concentrations were not toxic to strains assayed and did not affect biofilm formation (data not shown). After incubation for 48 h at 37 °C, the wells were washed three times with 200 µL of water to remove planktonic bacteria and dried. The remaining bacteria that adhered to the surface of the wells were fixed with 200 µL of methanol for 15 min. Then, the wells were emptied, air dried and subsequently stained with 200 µL of 2% crystal violet solution (Merck) for 5 min, followed by washing with water, drying and solubilizing the stain using 200 µL of 33% acetic acid for 10 min. The absorbance of each well was measured at 570 nm on a microplate reader (xMark Bio-Rad, Hercules, CA, USA). Additionally, CPX was subject to the same essay. The values are expressed as the percent biofilm inhibited in comparison to the untreated control biofilm (positive control). All tests were performed in triplicate.

4.4. Screening for Quorum Sensing Inhibition Activity

The QSI ability of the AKGs was evaluated in vitro in 96-well microtiter plates by a *C. violaceum* ATCC 12472 biosensor bioassay [4], as previously described with some modifications [41]. Briefly, preinoculum cultured in LB supplemented with kanamycin was adjusted to 1×10^9 CFU/mL at 600 nm. Then, 100 µL was added to each well, followed by 100 µL of AKGs dissolved in DMSO aqueous solution at subinhibitory concentrations of each compound or of the highest MIC assessed (250 µg/mL) and the plate was incubated for 48 h at 22 °C. Thus, the QSI activity was evaluated as the minimum quantity in µg/mL of sample required to inhibit violacein pigment and it was established by the appearance of a colorless and opaque well and did not have an effect on bacterial growth. The sample solvent that had no antibacterial activity served as the negative control and *C. violaceum* 12472 grown in the presence of the same amount of solvent without AKG was used as the positive control. To determine the effect of the presence of each AKG on the growth of the biosensor, a colony count was performed in each assay. All experiments were carried out with three replicates.

4.5. Statistical Analysis

All antibiofilm experiments were performed in triplicate and the results were expressed as the means ± standard deviation and calculated using statistical analyses of random uncertainties and rejection of data [65]. MIC and QSI data are representative of three independent experiments and the minimum concentration is expressed as the mean value.

5. Conclusions

AKGs are known bioactive compounds with weak to moderate antibacterial activity. Here, fifteen natural enantiopure AKGs were tested to determine their effect on biofilm development by seven clinical bacterial isolates and two reference strains as well their effect on the QSI activity in *C. violaceum* ATCC 12472 by microtiter plate assays. Ciprofloxacin as the standard antibiotic underwent the same antibiofilm assay. At subinhibitory concentrations, the highest biofilm inhibition rates (%) exhibited for all AKGs were influenced in a concentration-dependent manner and each AKG acted individually against the bacterial isolates, reaching rates up to 97.2%. In addition, AKGs displayed minimum QS inhibitory concentrations at different levels but did not affect the growth of *C. violaceum*. (2*S*)-3-*O*-(*cis*-13′-docosenyl)-1,2-propanediol **15** was the most effective AKG against QSI (20 µM), while (2*S*)-3-*O*-(*cis*-9′-hexadecenyl)-1,2-propanediol **7** was the least active (795 µM). Additional to the novelty of antibiofilm data, the results showed for the first time the QSI activity of this natural AKG series, which indeed, suggests that AKGs constitute a class of promising candidates for further studies on preventing antimicrobial resistance.

Author Contributions: K.E.C.D. performed the microbiological studies and interpreted the results. E.S.G. contributed to the experimental work supervision, data analysis and manuscript revision. B.D.M.M. reviewed the full paper. H.M.W. conducted the project supervision and manuscript preparation. All authors have read and agreed to the published version of the manuscript.

Funding: This research was funded by Universidad Nacional de Colombia, Vicerrectoría de Investigación, project number 46978.

Acknowledgments: The authors gratefully acknowledge financial support from the Universidad Nacional de Colombia. The authors express their gratitude to Catalina Arevalo Ferro from Universidad Nacional de Colombia, Grupo de Comunicación y Comunidades Bacterianas, Departamento de Biología, for her great assistance.

Conflicts of Interest: The authors declare that they had no conflict of interest.

References

1. Aydemir, D.H.; Cifci, G.; Aviyente, V.; Tinaz, G.B. Quorum-sensing inhibitor potential of trans-anethole against *Pseudomonas aeruginosa*. *J. Appl. Microbiol.* **2018**, *125*, 731–739. [CrossRef] [PubMed]
2. Chang, Y.; Wang, P.C.; Ma, H.M.; Chen, S.Y.; Fu, Y.H.; Liu, Y.Y.; Wang, X.; Yu, G.C.; Huang, T.; Hibbs, D.E.; et al. Design, synthesis and evaluation of halogenated furanone derivatives as quorum sensing inhibitors in *Pseudomonas aeruginosa*. *Eur. J. Pharm. Sci.* **2019**, *140*, 105058. [CrossRef] [PubMed]
3. Pacios, O.; Blasco, L.; Bleriot, I.; Fernandez, L.G.; González, M.B.; Ambroa, A.; López, M.; Bou, G.; Tomás, M. Strategies to combat multidrug-resistant and persistent infectious diseases. *Antibiotics* **2020**, *9*, 65. [CrossRef] [PubMed]
4. Reina, J.C.; Pérez, I.V.; Martín, J.; Llamas, I. A quorum-sensing inhibitor strain of *Vibrio alginolyticus* blocks Qs-controlled phenotypes in *Chromobacterium violaceum* and *Pseudomonas aeruginosa*. *Mar. Drugs* **2019**, *17*, 494. [CrossRef]
5. Fong, J.; Yuan, M.; Jakobsen, T.H.; Mortensen, K.T.; Delos Santos, M.M.; Chua, S.L.; Yang, L.; Tan, C.H.; Nielsen, T.E.; Givskov, M. Disulfide bond-containing ajoene analogs as novel quorum sensing inhibitors of Pseudomonas aeruginosa. *J. Med. Chem.* **2017**, *60*, 215–227. [CrossRef]
6. Xu, X.J.; Zeng, T.; Huang, Z.X.; Xu, X.F.; Lin, J.; Chen, W.M. Synthesis and biological evaluation of cajaninstilbene acid and amorfrutins A and B as inhibitors of the *Pseudomonas aeruginosa* quorum sensing system. *J. Nat. Prod.* **2018**, *81*, 2621–2629. [CrossRef]
7. Li, C.; Jiang, C.; Jing, H.; Jiang, C.; Wang, H.; Du, X.; Lou, Z. Separation of phenolics from peony flowers and their inhibitory activities and action mechanism on bacterial biofilm. *Appl. Microbiol. Biotechnol.* **2020**, *104*, 4321–4332. [CrossRef]
8. Rajkumari, J.; Borkotoky, S.; Murali, A.; Suchiang, K.; Mohanty, S.K.; Busi, S. Attenuation of quorum sensing controlled virulence factors and biofilm formation in *Pseudomonas aeruginosa* by pentacyclic triterpenes, betulin and betulinic acid. *Microb. Pathog.* **2018**, *118*, 48–60. [CrossRef]
9. Parasuraman, P.; Devadatha, B.; Sarma, V.V.; Ranganathan, S.; Ampasala, D.R.; Siddhardha, B. Anti-quorum sensing and antibiofilm activities of *Blastobotrys parvus* PPR3 against *Pseudomonas aeruginosa* PAO1. *Microb. Pathog.* **2020**, *138*, 103811. [CrossRef]
10. Vestby, L.K.; Grønseth, T.; Simm, R.; Nesse, L.L. Bacterial biofilm and its role in the pathogenesis of disease. *Antibiotics* **2020**, *9*, 59. [CrossRef]
11. Borges, A.; Sousa, P.; Gaspar, A.; Vilar, S.; Borges, F.; Simões, M. Furvina inhibits the 3-oxo-C12-HSL-based quorum sensing system of *Pseudomonas aeruginosa* and QS-dependent phenotypes. *Biofouling* **2017**, *33*, 156–168. [CrossRef]
12. Ohta, T.; Fukumoto, A.; Iizaka, Y.; Kato, F.; Koyama, Y.; Anzai, Y. Quorum sensing inhibitors against *Chromobacterium violaceum* CV026 derived from an actinomycete metabolite library. *Biol. Pharm. Bull.* **2020**, *43*, 179–183. [CrossRef]
13. Paguigan, N.D.; Rivera, J.C.; Stempin, J.J.; Augustinović, M.; Noras, A.I.; Raja, H.A.; Todd, D.A.; Triplett, K.D.; Day, C.; Figueroa, M.; et al. Prenylated diresorcinols inhibit bacterial quorum sensing. *J. Nat. Prod.* **2019**, *82*, 550–558. [CrossRef]
14. Chang, A.; Sun, S.; Li, L.; Dai, X.; Li, H.; He, Q.; Zhu, H. Tyrosol from marine fungi, a novel quorum sensing inhibitor against *Chromobacterium violaceum* and *Pseudomonas aeruginosa*. *Bioorg. Chem.* **2019**, *91*, 103140. [CrossRef]
15. Abraham, W.R. Going beyond the control of quorum-sensing to combat biofilm infections. *Antibiotics* **2016**, *5*, 3. [CrossRef]
16. Newman, D.J.; Cragg, G.M. Natural products as sources of new drugs over the nearly four decades from 01/1981 to 09/2019. *J. Nat. Prod.* **2020**, *83*, 770–803. [CrossRef]
17. Linthorne, J.S.; Chang, B.J.; Flematti, G.R.; Ghisalberti, E.L.; Sutton, D.C. A direct pre-screen for marine bacteria producing compounds inhibiting quorum sensing reveals diverse planktonic bacteria that are bioactive. *Mar. Biotechnol.* **2015**, *17*, 33–42. [CrossRef]
18. Carroll, A.R.; Copp, B.R.; Davis, R.A.; Keyzers, R.A.; Prinsep, M.R. Marine natural products. *Nat. Prod. Rep.* **2020**, *37*, 175–223. [CrossRef]
19. Deniau, A.L.; Mosset, P.; Pédrono, F.; Mitre, R.; Le Bot, D.; Legrand, A.B. Multiple beneficial health effects of natural alkylglycerols from shark liver oil. *Mar. Drugs* **2010**, *8*, 2175–2184. [CrossRef]
20. Magnusson, C.D.; Haraldsson, G.G. Ether lipids. *Chem. Phys. Lipids* **2011**, *164*, 315–340. [CrossRef]

21. Pemha, R.; Pegnyemb, D.E.; Mosset, P. Synthesis of (*Z*)-(20*R*)-1-*O*-(20-methoxynonadec-100-enyl)-sn-glycerol, a new analog of bioactive ether lipids. *Tetrahedron* **2012**, *68*, 2973–2983. [CrossRef]
22. Magnusson, C.D.; Gudmundsdottir, A.V.; Hansen, K.A.; Haraldsson, G.G. Synthesis of enantiopure reversed structured ether lipids of the 1-*O*-alkyl-sn-2,3-diacylglycerol type. *Mar. Drugs* **2015**, *13*, 173–201. [CrossRef] [PubMed]
23. Dean, J.M.; Lodhi, I.J. Structural and functional roles of ether lipids. *Protein Cell* **2018**, *9*, 196–206. [CrossRef] [PubMed]
24. Iannitti, T.; Palmieri, B. An update on the therapeutic role of alkylglycerols. *Mar. Drugs* **2010**, *8*, 2267–2300. [CrossRef]
25. Watschinger, K.; Werner, E.R. Orphan enzymes in ether lipid metabolism. *Biochimie* **2013**, *95*, 59–65. [CrossRef]
26. Sutter, M.; Silva, E.D.; Duguet, N.; Raoul, Y.; Métay, E.; Lemaire, M. Glycerol ether synthesis: A bench test for green chemistry concepts and technologies. *Chem. Rev.* **2015**, *115*, 8609–8651. [CrossRef]
27. Tyrtyshnaia, A.; Manzhulo, I.; Kipryushina, Y.; Ermolenko, E. Neuroinflammation and adult hipocampal neurogenesis in neuropathic pain and alkyl glycerol ethers treatment in aged mice. *Int. J. Mol. Med.* **2019**, *43*, 2153–2163. [CrossRef]
28. Yu, H.; Dilbaz, S.; Coßmann, J.; Hoang, A.C.; Diedrich, V.; Herwig, A.; Harauma, A.; Hoshi, Y.; Moriguchi, T.; Landgraf, K.; et al. Breast milk alkylglycerols sustain beige adipocytes through adipose tissue macrophages. *J. Clin. Investig.* **2019**, *129*, 2485–2499. [CrossRef]
29. Haynes, M.P.; Buckley, H.R.; Higgins, M.L.; Pieringer, R.A. Synergism between the antifungal agents amphotericin B and alkylglycerol ethers. *Antimicrob. Agents Chemother.* **1994**, *38*, 1523–1529. [CrossRef]
30. Ved, H.S.; Gustow, E.; Mahadevans, V.; Pieringer, R.A. Dodecylglycerol a new type of antibacterial agent which stimulates autolysin activity in *Streptococcus faecium* ATCC 9790. *J. Biol. Chem.* **1984**, *259*, 8115–8121. [CrossRef]
31. Brissette, J.L.; Cabacungan, E.A.; Pieringer, R.A. Studies on the antibacterial activity of dodecylglycerol. Its limited metabolism and inhibition of glycerolipid and lipoteichoic acid biosynthesis in *Streptococcus mutans* BHT. *J. Biol. Chem.* **1986**, *261*, 6338–6345. [CrossRef]
32. Ved, H.S.; Gustow, E.; Pieringer, R.A. Synergism between penicillin G and the antimicrobial ether lipid, *rac*-l-dodecylglycerol, acting below its critical micelle concentration. *Lipids* **1990**, *25*, 119–121. [CrossRef]
33. Lampe, M.F.; Ballweber, L.M.; Isaacs, C.E.; Patton, D.L.; Stamm, W.E. Killing of *Chlamydia trachomatis* by novel antimicrobial lipids adapted from compounds in human breast milk. *Antimicrob. Agents Chemother.* **1998**, *42*, 1239–1244. [CrossRef]
34. Qi, S.H.; Zhang, S.; Yang, L.H.; Qian, P.Y. Antifouling and antibacterial compounds from the gorgonians *Subergorgia suberosa* and *Scirpearia gracilis*. *Nat. Prod. Rep.* **2008**, *22*, 154–166. [CrossRef]
35. Lin, Y.C.; Schlievert, P.M.; Anderson, M.J.; Fair, C.L.; Schaefers, M.M.; Muthyala, R.; Peterson, M.L. Glycerol monolaurate and dodecylglycerol effects on *Staphylococcus aureus* and toxic shock syndrome toxin-1 in vitro and in vivo. *PLoS ONE* **2009**, *4*, e7499. [CrossRef]
36. Castellanos, L.H.; Mayorga, H.W.; Duque, C.B. Estudio de la composición química y actividad *antifouling* del extracto de la esponja marina *Cliona delitrix*. *Vitae* **2010**, *17*, 209–224.
37. Mayorga, H.; Urrego, N.F.; Castellanos, L.; Duque, C. Cembradienes from the Caribbean Sea whip *Eunicea* sp. *Tetrahedron Lett.* **2011**, *52*, 2515–2518. [CrossRef]
38. Martínez, Y.D.; Vanegas, G.L.; Reina, L.G.; Mayorga, H.W.; Arévalo-Ferro, C.; Ramos, F.R.; Duque, C.B.; Castellanos, L.H. Biofilm inhibition activity of compounds isolated from two *Eunicea* species collected at the Caribbean Sea. *Rev. Bras. Farmacogn.* **2015**, *25*, 605–611. [CrossRef]
39. Othmani, A.; Bunet, R.; Bonnefont, J.L.; Briand, J.F.; Culioli, G. Settlement inhibition of marine biofilm bacteria and barnacle larvae by compounds isolated from the Mediterranean brown alga *Taonia atomaria*. *J. Appl. Phycol.* **2015**, *28*, 1975–1986. [CrossRef]
40. Nascimento, T.S.; Monteiro, L.G.; Braga, E.F.; Batista, W.R.; Albert, A.L.; Chantre, L.G.; Machado, S.P.; Lopes, R.S.; Lopes, C.C. Synthesis of natural ether lipids and 1-*O*-hexadecylglycero-arylboronates via an epoxidering opening approach: Potential antifouling additives to marine paint coatings. *Int. J. Adv. Res. Sci. Eng. Technol.* **2018**, *5*, 326–332. [CrossRef]
41. Barragán, C.A.; Silva, E.G.; Moreno, B.M.; Mayorga, H.W. Inhibition of quorum sensing by compounds from two *Eunicea* species and synthetic saturated alkylglycerols. *Vitae* **2018**, *25*, 92–103. [CrossRef]
42. Fernández, D.M.; Contreras, L.J.; Moreno, B.M.; Silva, E.G.; Mayorga, H.W. Enantiomeric synthesis of natural alkylglycerols and their antibacterial and antibiofilm activities. *Nat. Prod. Res.* **2019**, 1–7. [CrossRef]
43. Qiu, M.N.; Wang, F.; Chen, S.Y.; Wang, P.C.; Fu, Y.H.; Liu, Y.Y.; Wang, X.; Wang, F.B.; Wang, C.; Yang, H.W.; et al. Novel 2, 8-bit derivatives of quinolines attenuate *Pseudomonas aeruginosa* virulence and biofilm formation. *Bioorg. Med. Chem. Lett.* **2019**, *29*, 749–754. [CrossRef]
44. Kumar, P.; Lee, J.H.; Beyenal, H.; Lee, J. Fatty acids as antibiofilm and antivirulence agents. *Trends Microbiol.* **2020**, *28*, 753–768. [CrossRef]
45. Rabin, N.; Zheng, Y.; Opoku, C.T.; Du, Y.; Bonsu, E.; Sintim, H.O. Agents that inhibit bacterial biofilm formation. *Future Med. Chem.* **2015**, *7*, 647–671. [CrossRef]
46. Nunes, L.S.; Rigon, Z.K.; Macedo, A.J.; Silva, D.T. Plant natural products targeting bacterial virulence factors. *Chem. Rev.* **2016**, *116*, 9162–9236. [CrossRef]
47. Townsley, L.; Shank, E.A. Natural-product antibiotics: Cues for modulating bacterial biofilm formation. *Trends Microbiol.* **2017**, *25*, 1016–1026. [CrossRef]
48. Song, X.; Xia, Y.X.; He, Z.D.; Zhang, H.J. A review of natural products with anti-biofilm activity. *Curr. Org. Chem.* **2018**, *22*, 788–816. [CrossRef]

49. Melander, R.J.; Basak, A.K.; Melander, C. Natural products as inspiration for the development of bacterial antibiofilm agents. *Nat. Prod. Rep.* **2020**, *37*, 1454–1477. [CrossRef]
50. López, Y.; Soto, S.M. The usefulness of microalgae compounds for preventing biofilm infections. *Antibiotics* **2020**, *9*, 9. [CrossRef]
51. Asfour, H.Z. Anti-quorum sensing natural compounds. *J. Microsc. Ultrastruct.* **2018**, *6*, 1–10. [CrossRef] [PubMed]
52. Deryabin, D.; Galadzhieva, A.; Kosyan, D.; Duskaev, G. Plant-derived inhibitors of AHL-mediated quorum sensing in bacteria: Modes of action. *Int. J. Mol. Sci.* **2019**, *20*, 5588. [CrossRef] [PubMed]
53. Saurav, K.; Costantino, V.; Venturi, V.; Steindler, L. Quorum sensing inhibitors from the sea discovered using bacterial *N*-acyl-homoserine lactone-based biosensors. *Mar. Drugs* **2017**, *15*, 53. [CrossRef] [PubMed]
54. Zhao, J.; Li, X.; Hou, X.; Quan, C.; Chen, M. Widespread existence of quorum sensing inhibitors in marine bacteria: Potential drugs to combat pathogens with novel strategies. *Mar. Drugs* **2019**, *17*, 275. [CrossRef]
55. Cadavid, E.; Echeverri, F. The search for natural inhibitors of biofilm formation and the activity of the autoinductor C6-AHL in *Klebsiella pneumoniae* ATCC 13884. *Biomolecules* **2019**, *9*, 49. [CrossRef]
56. Dos Reis Ponce, A.; Martin, M.L.; De Araujo, E.F.; Mantovani, H.L.; Vanetti, M.C. AiiA quorum-sensing quenching controls proteolytic activity and biofilm formation by Enterobacter cloacae. *Curr. Microbiol.* **2012**, *65*, 758–763. [CrossRef]
57. Qvortrup, K.; Hultqvist, L.D.; Nilsson, M.; Jakobsen, T.H.; Jansen, C.U.; Uhd, J.; Andersen, J.B.; Nielsen, T.E.; Givskov, M.; Nielsen, T.T. Small molecule anti-biofilm agents developed on the basis of mechanistic understanding of biofilm formation. *Front. Chem.* **2019**, *7*, 742. [CrossRef]
58. Wasfi, R.; Hamed, S.M.; Amer, M.A.; Fahmy, L.I. *Proteus mirabilis* biofilm: Development and therapeutic strategies. *Front. Cell. Infect. Microbiol.* **2020**, *10*, 414. [CrossRef]
59. Wang, H.; Chu, W.; Ye, C.; Gaeta, B.; Tao, H.; Wang, M.; Qiu, Z. Chlorogenic acid attenuates virulence factors and pathogenicity of *Pseudomonas aeruginosa* by regulating quorum sensing. *Appl. Microbiol. Biotechnol.* **2019**, *103*, 903–915. [CrossRef]
60. Li, Y.; Pan, J.; Wu, D.; Tian, Y.; Zhang, J.; Fang, J. Regulation of *Enterococcus faecalis* biofilm formation and quorum sensing related virulence factors with ultra-low dose reactive species produced by plasma activated water. *Plasma Chem. Plasma Process.* **2019**, *39*, 35–49. [CrossRef]
61. Monticelli, J.; Knezevich, A.; Luzzati, R.; Di Bella, S. Clinical management of non-*faecium* non-*faecalis* vancomycin-resistant enterococci infection. Focus on *Enterococcus gallinarum* and *Enterococcus casseliflavus/flavescens*. *J. Infect. Chemother.* **2018**, *24*, 237–246. [CrossRef]
62. Lönn, J.S.; Naemi, A.O.; Benneche, T.; Petersen, F.C.; Scheie, A.A. Thiophenones inhibit *Staphylococcus epidermidis* biofilm formation at nontoxic concentrations. *FEMS Immunol. Med. Microbiol.* **2012**, *65*, 326–334. [CrossRef]
63. Nakagawa, S.; Hillebrand, G.G.; Nunez, G. *Rosmarinus officinalis* L. (Rosemary) extracts containing carnosic acid and carnosol are potent quorum sensing inhibitors of *Staphylococcus aureus* Virulence. *Antibiotics* **2020**, *9*, 149. [CrossRef]
64. Stepanović, S.; Vuković, D.; Dakić, I.; Savić, B.; Svabić-Vlahovic, M. A modified microtiter-plate test for quantification of staphylococcal biofilm formation. *J. Microbiol. Methods* **2000**, *40*, 175–179. [CrossRef]
65. Taylor, J.R. *An Introduction to Error Analysis. The Study of Uncertainties in Physical Measurements*, 2nd ed.; University Science Books: Sausalito, CA, USA, 1997; pp. 93–170.

 MDPI

Article
Antifungal and Anti-Biofilm Effects of Caffeic Acid Phenethyl Ester on Different *Candida* Species

Ibrahim Alfarrayeh [1,2,*], Edit Pollák [3], Árpád Czéh [4], András Vida [4], Sourav Das [5] and Gábor Papp [1]

1 Department of General and Environmental Microbiology, Faculty of Science, University of Pécs, Ifjúság Str. 6, 7624 Pécs, Hungary; pappgab@gamma.ttk.pte.hu
2 Department of Biological Sciences, Faculty of Science, Mu'tah University, Mu'tah University Street, Al-Karak 61710, Jordan
3 Department of Animal Anatomy and Developmental Biology, Faculty of Science, University of Pécs, Ifjúság Str. 6, 7624 Pécs, Hungary; peditmail@gmail.com
4 Soft Flow Hungary R&D Ltd., Ürögi Fasor Str. 2/A, 7634 Pécs, Hungary; aczeh@foss.dk (Á.C.); avida@foss.dk (A.V.)
5 Department of Laboratory Medicine, Medical School, University of Pécs, Ifjúság Str. 13, 7624 Pécs, Hungary; pharma.souravdas@gmail.com
* Correspondence: alfarrayeh@gmail.com; Tel.: +962-790-632-435

Abstract: This study investigated the effect of CAPE on planktonic growth, biofilm-forming abilities, mature biofilms, and cell death of *C. albicans*, *C. tropicalis*, *C. glabrata*, and *C. parapsilosis* strains. Our results showed a strain- and dose-dependent effect of CAPE on *Candida*, and the MIC values were between 12.5 and 100 µg/mL. Similarly, the MBIC values of CAPE ranging between 50 and 100 µg/mL highlighted the inhibition of the biofilm-forming abilities in a dose-dependent manner, as well. However, CAPE showed a weak to moderate biofilm eradication ability (19-49%) on different *Candida* strains mature biofilms. Both caspase-dependent and caspase-independent apoptosis after CAPE treatment were observed in certain tested *Candida* strains. Our study has displayed typical apoptotic hallmarks of CAPE-induced chromatin margination, nuclear blebs, nuclear condensation, plasma membrane detachment, enlarged lysosomes, cytoplasm fragmentation, cell wall distortion, whole-cell shrinkage, and necrosis. In conclusion, CAPE has a concentration and strain-dependent inhibitory activity on viability, biofilm formation ability, and cell death response in the different *Candida* species.

Keywords: CAPE; *Candida*; antifungal; biofilm; apoptosis

Citation: Alfarrayeh, I.; Pollák, E.; Czéh, Á.; Vida, A.; Das, S.; Papp, G. Antifungal and Anti-Biofilm Effects of Caffeic Acid Phenethyl Ester on Different *Candida* Species. *Antibiotics* **2021**, *10*, 1359. https://doi.org/10.3390/antibiotics10111359

Academic Editors: Luís Melo and Andreia S. Azevedo

Received: 30 September 2021
Accepted: 3 November 2021
Published: 7 November 2021

1. Introduction

The genus *Candida* refers to a yeast that is part of the microbiota of healthy individuals living in commensalism with the human. However, in some cases, few *Candida* species tend to be opportunistic fungal pathogens, causing candidiasis. Candidiasis is among the common human infections; its symptoms vary according to the location of the infection in the body. Most of the infections may lead to minor symptoms such as slight localized rashes, redness, itching, and discomfort, though symptoms can be severe or even fatal if left without treatment in immunocompromised individuals [1–3]. Mostly, candidiasis is attributed to *Candida albicans*, however; non-albicans *Candida* species, including *C. parapsilosis*, *C. tropicalis*, and *C. glabrata*, have been reported to cause 30% to 54% of *Candida* infections [4–6]. Furthermore, the ability of these species to exhibit multidrug resistance, which may cause failure of the antifungal therapy, has been reported in earlier studies [6].

The ability of *Candida* spp to shift the commensal to pathogenic lifestyle is attributed to the presence of several virulence factors. Predominantly, the capability of switching morphology between yeast and hyphal forms, and the ability to form biofilms are the major properties crucial to *Candida* spp pathogenesis. The *Candida* infections are accompanied by the formation of biofilms on host tissues, organs, or abiotic surfaces such as urinary catheters, which result in high morbidity and mortality [7–9]. As with all microbial biofilms, *Candida* biofilms are highly resistant to antimicrobial treatment. Several factors might contribute to this resistance, including the physiological state of *Candida* cells in the biofilm, extracellular polymeric substances, overexpression of membrane-localized drug efflux pumps, variations in sterol content in the fungal membrane, and different developmental phases of cells through the biofilm [10]. Thus, the effectiveness of the current therapeutic agents against *Candida* biofilms is considered low, with a few exceptions [7–9]. The biofilms of *Candida* were 30 to 2000 times more resistant than planktonic cells to many antifungal agents, including amphotericin B, fluconazole, itraconazole, and ketoconazole [11]. As a result, the need to provide natural alternatives to these synthetic antifungal agents has arisen. The antifungal mechanisms of action of these natural alternatives can include the inhibition of germination and biofilm formation, the disruption of cell wall integrity, the alteration of cell membrane permeability, or the induction of apoptosis [9].

Caffeic acid phenethyl ester (CAPE), also called phenylethyl caffeate or phenethyl caffeate, is one of the promising natural alternatives of synthetic antimicrobial drugs. This polyphenolic ester compound is a major component of temperate propolis (poplar-type) and can be produced in the laboratory by reacting caffeic acid with phenethyl alcohols. It consists of hydroxyl groups within the catechol ring, which is crucial for many biological activities [12,13]. No information was available in the literature about the LD_{50} of CAPE in animal models or on normal human cells. However, Koru and coworkers investigated cytotoxicity in the human multiple myeloma cell line with LD_{50} at 24, 48, and 72 h, found at 49.1, 30.6, and 22.5 μg/mL, respectively [14]. The CAPE has a wide range of biological activities, such as inhibition of nuclear factor κ-B, cell division restriction, termination of the cell cycle, and apoptotic induction [15]. It has been studied extensively as the most important individual component of propolis. Existing studies focused on its potential therapeutic properties such as its antibiotic, antioxidant, anti-inflammatory, anti-oxidative stress, antitumor, antidiabetic, anti-neurodegeneration, and anti-anxiety properties [13,16,17]. Numerous studies demonstrated the antibacterial activity of CAPE against different bacterial species [18–22]. However, few studies that have investigated the antifungal activity of CAPE as a single molecule or in combination with some antibiotics were found in the literature. De Barros and coworkers recently reported the ability of CAPE to inhibit the growth of both fluconazole-sensitive and fluconazole-resistant strains of *C. albicans* [23]. Moreover, Sun and coworkers found the synergistic effects of CAPE with caspofungin and fluconazole against *C. albicans* and fluconazole-resistant *C. albicans*, respectively [24,25]. The synergism with caspofungin was associated with a loss in iron homeostasis induced by CAPE, leading to functional defects in the mitochondrial respiratory chain and energy depletion, which increases the susceptibility of *C. albicans* to caspofungin [25].

Presently, CAPE has been given close attention for its important therapeutic effects in many diseases, including carcinomas, internal organ damage, metabolic diseases, inflammatory diseases, and microbial infections. CAPE may be a promising natural product for clinical application in the future [21]. One of the major advantages is that CAPE is devoid of some negative aspects of crude extracts of propolis, which includes, the inability to be standardized, which is a keystone of implementing it as therapy in the field of medicine [16]. This study aimed to investigate how CAPE, as a single molecule, affects planktonic growth, biofilm-forming abilities, mature biofilms, and cell death of some *Candida albicans* and non-albicans *Candida* species and strains.

2. Results

2.1. Susceptibility of *Candida* Planktonic Cells to CAPE

The antifungal effect of CAPE on nine *Candida* strains has been studied. The results of the minimal inhibitory concentration (MIC_{80}) values for CAPE against different *Candida* species and strains are given in Table 1. It has been found that CAPE has a strain- and dose-dependent effect. The MIC_{80} values ranged from 12.5 to 100 μg/mL. The highest inhibitory effect was seen against *C. glabrata* SZMC 1378, *C. glabrata* SZMC 1374, and *C. parapsilosis* SZMC 8008 compared with the other strains. However, the most resistant strain was *C. albicans* SZMC 1423.

Table 1. MIC_{80} values of CAPE against the different *Candida* strains.

Strain	MIC_{80} (μg/mL)
C. albicans ATCC 44829	50
C. albicans SZMC 1423	100
C. albicans SZMC 1424	50
C. glabrata SZMC 1374	12.5
C. glabrata SZMC 1378	12.5
C. parapsilosis SZMC 8007	25
C. parapsilosis SZMC 8008	12.5
C. tropicalis SZMC 1366	50
C. tropicalis SZMC 1512	50

2.2. *Effect of CAPE on Candida Biofilm-Forming Ability*

The biofilm-forming ability is a crucial property related to the pathogenicity of *Candida*. In this experiment, the effect of CAPE was tested on the biofilms of four different *Candida* strains with high biofilm-forming abilities. The results demonstrate that CAPE has a dose-dependent inhibitory effect on the biofilm formation in four strains (Figure 1). The minimal biofilm inhibitory concentration (MBIC) values were 50, 50, 50, and 100 μg/mL for *C. albicans* SZMC 1424, *C. glabrata* SZMC 1374, *C. parapsilosis* SZMC 8007, and *C. tropicalis* SZMC 1366, respectively.

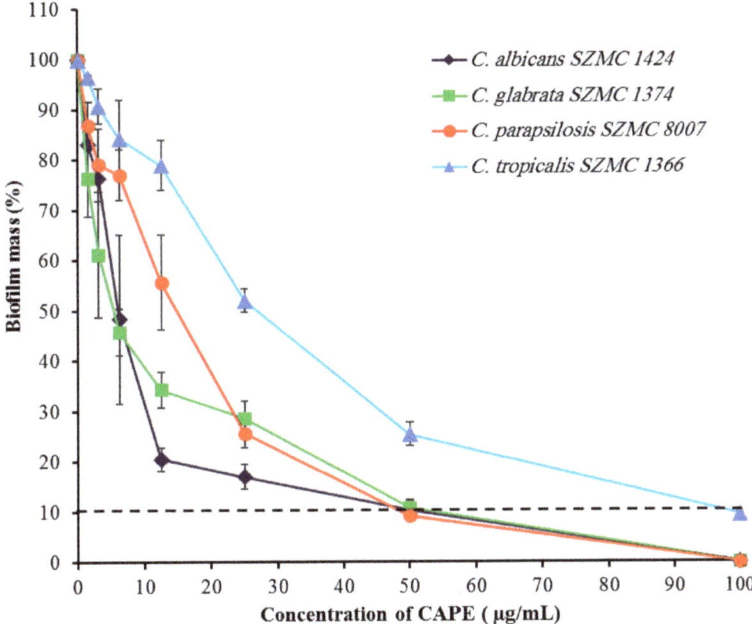

Figure 1. Effect of CAPE on the biofilm-forming ability of four *Candida* species. The dashed line represents the MBIC. Data are shown as mean ± SD from three independent experiments.

2.3. Effect of CAPE on Candida Biofilm Eradication

The effect of CAPE on the mature biofilms of four biofilm-forming *Candida* species was investigated. Treatment of the mature biofilms of *C. albicans* SZMC 1424, *C. glabrata* SZMC 1374, *C. tropicalis* SZMC 1366, and *C. parapsilosis* SZMC 8007 with different concentrations of CAPE caused a partial eradication. The maximum eradications (19–49%) for the mature biofilms of *C. glabrata* SZMC 1374, *C. albicans* SZMC 1424, and *C. parapsilosis* SZMC 8007 were achieved at 25, 50, and 50 μg/mL, respectively. Moreover, the eradication process was found to be dose-independent above 50 μg/mL in the case of these strains. On the other hand, the mature biofilms of *C. tropicalis* SZMC 1366 were the most resistant to CAPE, and the maximum eradication was achieved at 100 μg/mL (Figure 2).

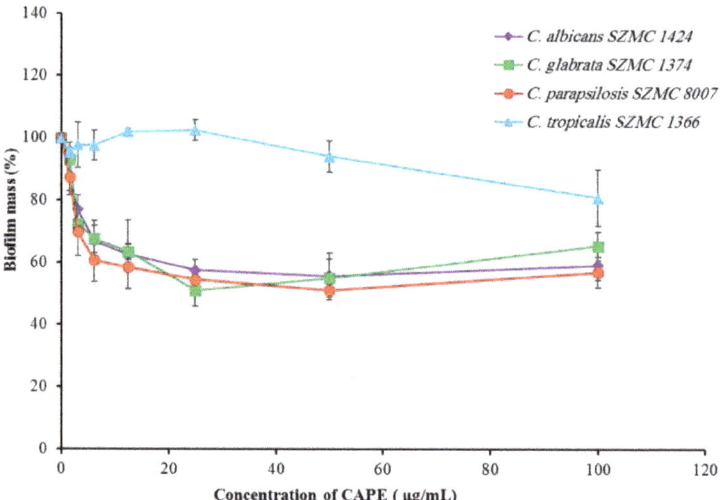

Figure 2. Effect of CAPE on mature biofilms of four *Candida* species. Data are shown as mean ± SD from three independent experiments.

2.4. Biosorption of CAPE by Candida Cells

Biosorption may be defined as "the removal/binding of desired substances from aqueous solution by biological material" [26]. The results revealed that the biosorption of CAPE by different *Candida* strains occurs rapidly, followed by a maximum biosorption observed within the first 30 to 90 min (Figure 3). According to the amount of CAPE biosorbed, two groups can be recognized: the first group was able to biosorb 53–63 μg/mL of CAPE, and it includes *C. albicans* SZMC 1424, *C. parapsilosis* SZMC 8007, and *C. parapsilosis* SZMC 8008; in contrast, the second group was able to biosorb 74–86 μg/mL, and it includes *C. albicans* ATCC 44829, *C. albicans* SZMC 1423, *C. tropicalis* SZMC 1366, *C. tropicalis* SZMC 1512, *C. glabrata* SZMC 1374, and *C. glabrata* SZMC 1378.

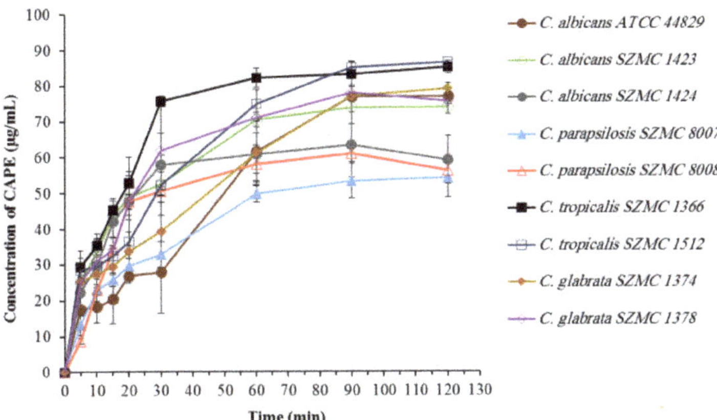

Figure 3. Cellular uptake of CAPE by nine *Candida* strains. Data are shown as mean ± SD from three independent experiments.

2.5. Induction of Apoptotic Cell Death in Candida spp. by CAPE

Cells of the nine *Candida* strains treated with sub-lethal concentrations of CAPE were analyzed by double staining with CF®488A Annexin V and propidium iodide (PI). The apoptotic cells with externalized phosphatidylserine were detected by CF®488A Annexin V, while necrotic cells were detected by PI staining. The results shown in Figure 4 demonstrate CAPE-induced apoptosis in six of the tested strains at different levels. Among these strains, *C. albicans* ATCC 44829 and *C. albicans* SZMC 1423 revealed the highest percentage of early apoptotic cells (69.8 and 70.2%, respectively), whereas almost no apoptosis was seen in *C. glabrata* SZMC 1374, *C. parapsilosis* SZMC 8008, and *C. glabrata* SZMC 1378 (apoptotic cells $\leq$ 2%). On the other hand, no necrosis was observed in any of the tested strains (necrotic cells $\leq$ 1%). Examples of the scatter plots can be found in the supplementary material (Figures S1–S7).

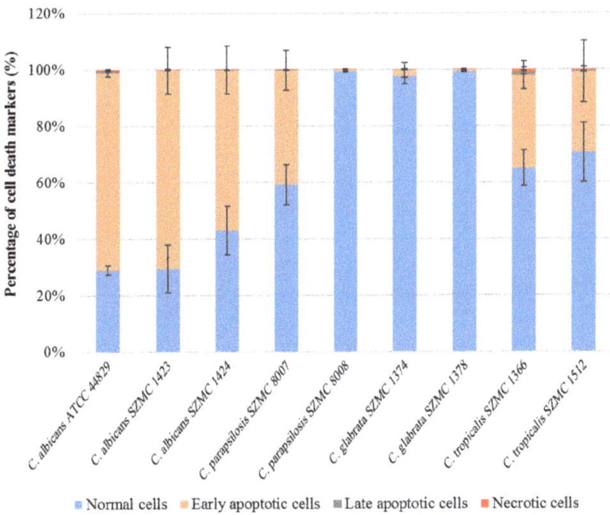

Figure 4. Cell death induced by CAPE treatment in nine *Candida* strains as determined by annexin V and PI staining. Data are shown as mean ± SD from three independent experiments.

2.6. Effect of Caspase Inhibitor on the Growth of CAPE-Treated Candida Cells

To investigate whether yeast caspase Yca1p is involved in CAPE-induced apoptotic cell death, pre-incubation with the pan-caspase inhibitor Z-VAD-FMK was applied for 1 h. The growth of the sub-lethal CAPE concentration-treated *Candida* strains that had apoptosis was analyzed with and without pre-incubation with Z-VAD-FMK. As shown in Figure 5, a significant increase in the viability was observed in *C. albicans* ATCC 44829, *C. albicans* SZMC 1424, *C. tropicalis* SZMC 1366, and *C. tropicalis* SZMC 1512 that are pre-incubated with the pan-caspase inhibitor Z-VAD-FMK. However, the viability of CAPE-treated *C. albicans* SZMC 1423 and *C. parapsilosis* SZMC 8007 was not affected by the pre-incubation with the pan-caspase inhibitor Z-VAD-FMK.

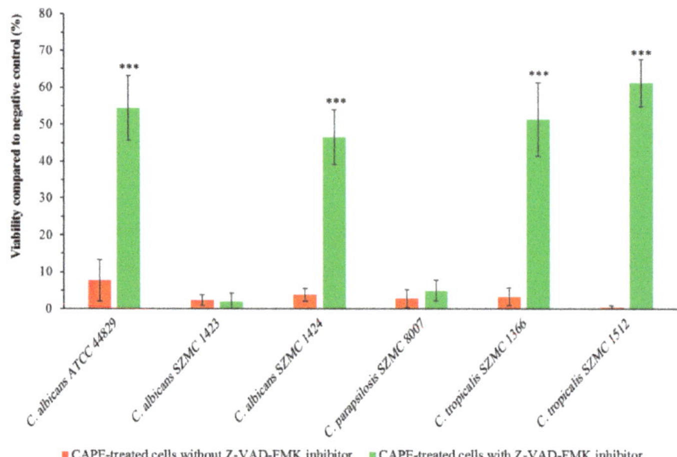

Figure 5. Effect of the pan-caspase inhibitor Z-VAD-FMK on the viability of six *Candida* strains treated with sub-lethal concentrations of CAPE. Data are shown as mean ± SD from three independent experiments. *** $p < 0.001$ indicates a significant increment of the viability compared with the viability without pre-incubation with the pan-caspase inhibitor Z-VAD-FMK.

2.7. CAPE-Induced Subcellular Cell Death Markers Determined by TEM

To visualize the changes in intracellular morphology of the cells after CAPE treatment, transmission electron microscopy imaging was performed on *Candida* cells exposed to sub-lethal concentrations of CAPE. The TEM micrographs of *C. tropicalis* SZMC 1366, *C. albicans* SZMC 1423, and *C. parapsilosis* SZMC 8007 (Figures 6–8, respectively) mainly revealed typical hallmarks of apoptosis, including nuclear chromatin margination, nuclear blebs, condensation in the nucleus, vacuolization, plasma membrane detachment, enlarged lysosomes, cytoplasm fragmentation, cell wall distortion, and whole-cell shrinkage. However, very few cells displayed signs of necrosis, such as membrane disintegration and loss of cytoplasm density, whereas the TEM micrographs of *C. glabrata* SZMC 1374 (Figure 9) mainly revealed smaller necrotic signs.

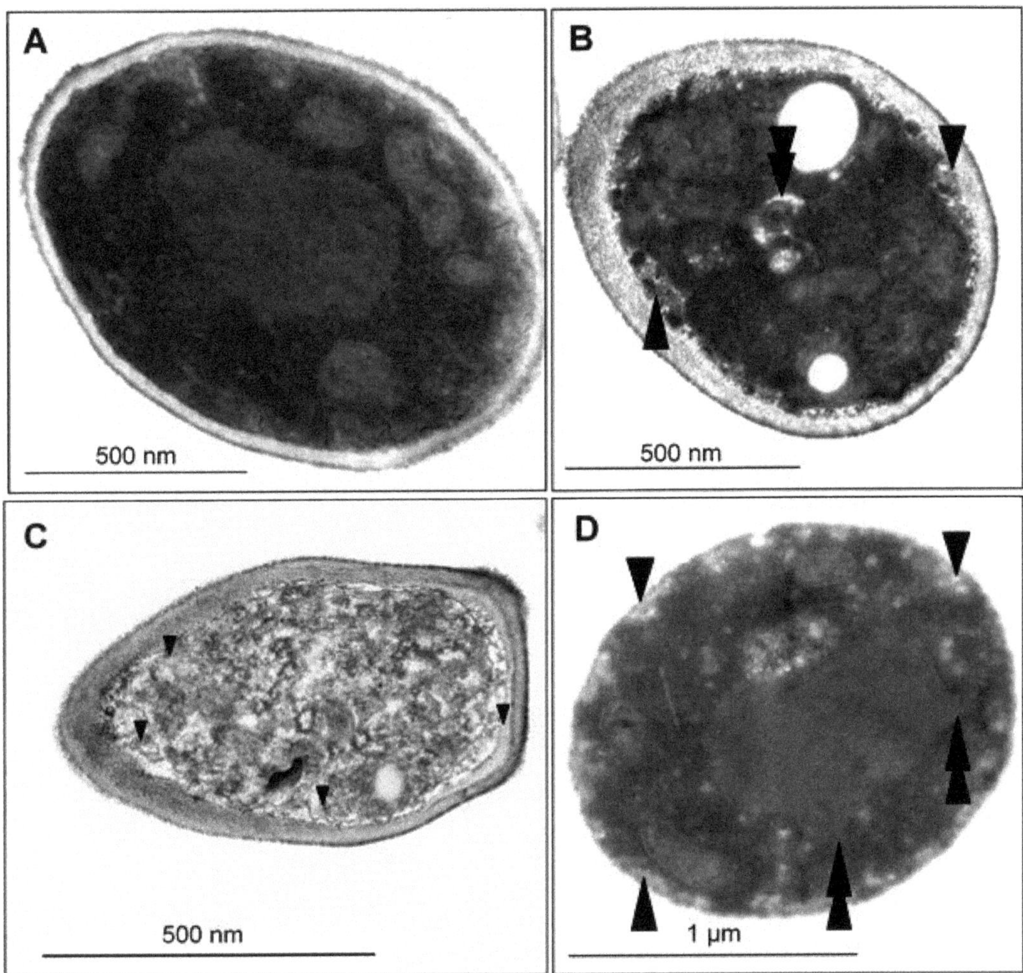

Figure 6. (**A**) TEM micrographs of the control *Candida tropicalis* SZMC 1366 cell structure demonstrates intact membranes, small unevenly scattered condensed chromatin grains, and homogenous cytoplasm structure. (**B–D**) TEM micrographs of *C. tropicalis* treated with a sub-lethal concentration of CAPE: (**B**) Late-stage disintegration with membrane fingerprints, vacuolization, plasma membrane detachment (arrowheads), and vacuole formation (double arrowheads). Fine granular homogenous cytoplasm organization disappeared, and a dense, compact cytoplasm with signs of fragmentation, rounded cell shape, and whole-cell shrinkage was seen. (**C**) Necrotic cell with membrane ruptures (arrowheads) and loss of cytoplasm density. (**D**) Several peripheral vacuoles show plasma membrane involvement (arrowheads). Nuclear bleb formation (double arrowheads).

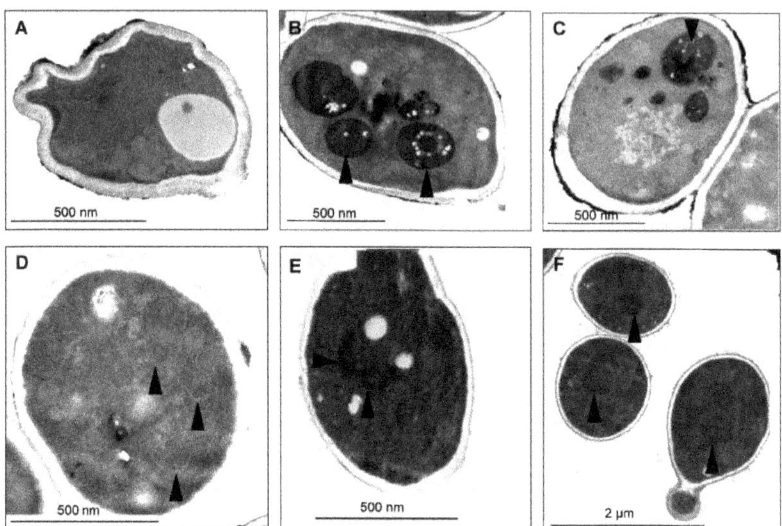

Figure 7. TEM micrographs of *C. albicans* SZMC 1423 treated with a sub-lethal concentration of CAPE exhibit different markers of cellular deterioration. (**A**) Severe cell wall distortion. (**B**,**C**) Appearance of enlarged lysosomes was rather frequently detected (arrowheads). (**D**) Isolation membranes precondition of cytoplasm fragmentation (arrowheads). (**E**,**F**) Nucleus fragmentation and marginal condensation (arrowheads).

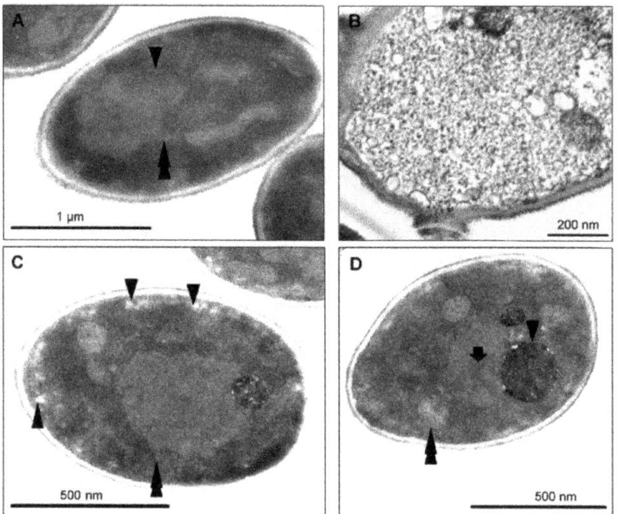

Figure 8. TEM micrographs of *C. parapsilosis* SZMC 8007 treated with a sub-lethal concentration of CAPE. Both apoptotic and necrotic cell structural changes were observed in samples. (**A**) Nuclear chromatin margination and condensation (arrowhead) and blebs (double arrowhead) detached from the nucleus are typical apoptotic hallmarks. (**B**) Few necrotic cells were also present. Note membrane disintegration, obvious vacuolization, and loss of cytoplasm density. (**C**) Peripheral vacuole formation refers to Golgi fragmentation and cell membrane separation from the cell wall (arrowheads). Nuclear blebs (double arrowhead). (**D**) Nuclear condensation (arrow) and extremely large lysosomal bodies (arrowhead). Note the swollen mitochondria (double arrowheads).

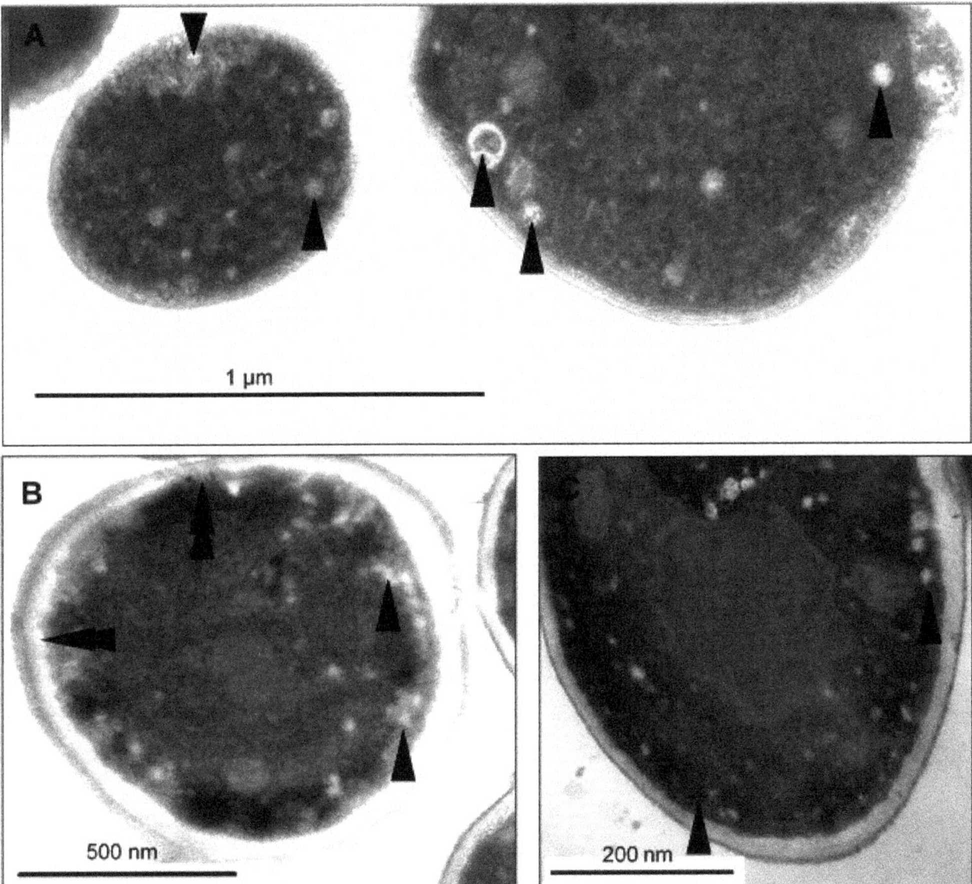

Figure 9. TEM micrographs of *C. glabrata* SZMC 1374 treated with a sub-lethal concentration of CAPE. (**A–C**) Mainly smaller necrotic signs were detected. Arrowheads denote small, mainly peripheral vacuoles typical of all samples. (**B**) Cell wall disintegration was also observed (double arrowhead).

3. Discussion

This study focused on the antifungal and anti-biofilm effects of CAPE, which is one of the main biologically active components of propolis, on different *Candida* species including *C. albicans* and non-albicans *Candida* species. Moreover, we also investigated some of the mechanisms that might be involved in CAPE-induced cell death.

The *Candida* spp are still considered the most important opportunistic fungal pathogens that cause fungal infections worldwide. They are among the fourth to sixth most common nosocomial bloodstream isolates, according to estimates. Although *C. albicans* was the most frequent species isolated during candidemia, a greater role of non-albicans *Candida* spp has been observed in recent years [6,27]. Moreover, the extensive use of azole antifungals has resulted in the development of multidrug resistance in many *Candida* strains [28,29]. This resistance could be attributed to a change in drug intracellular accumulation, a change in membrane sterol composition, a change in efflux pump performance, or a change in ERG11 (the gene that is responsible for the production of the lanosterol-14-demethylase enzyme, the target of these medications) [6].

The biological activities of CAPE have been widely studied. However, limited number of studies were found in the literature concerning the antifungal activity. This study aimed to add some information about the activity of CAPE against *C. albicans* and non-albicans *Candida* species. Our findings showed that CAPE has high abilities to inhibit planktonic growth and biofilm formation as well as an ability to partially eradicate the mature biofilms of the different strains of *Candida*. Those results are in agreement with the results obtained by De Barros and coworkers, where the ability of CAPE to inhibit the growth of both fluconazole-sensitive and fluconazole-resistant strains of *C. albicans* has been reported [23]. Another study performed by Possamai Rossatto and coworkers showed similar results, where CAPE was also able to inhibit the growth of *C. auris* with MIC values ranging from 1 to 64 μg/mL. Furthermore, CAPE was able to inhibit the biofilm formation and phospholipase production of *C. auris* [30]. Our findings also showed the ability of CAPE to enter the cells of *Candida* spp rapidly. Such results were also observed by Cigut and coworkers in the yeast *Saccharomyces cerevisiae*. They found that, out of the four examined compounds (caffeic acid, p-coumaric acid, ferulic acid, and CAPE), CAPE was the only one that was able to enter the *S. cerevisiae* cells [31].

Phenolic compounds, including CAPE, are effective inhibitors of iron absorption [25]. According to Sun and coworkers, the antifungal mechanism of CAPE may include intracellular iron starvation due to its ability to form insoluble complexes with iron ions, which leads to the prevention of iron absorption by cells [25]. Using *C. albicans*-infected nematodes, Breger and coworkers showed that CAPE was able to inhibit the in vivo filamentation of *C. albicans*, leading to prolonged survival of infected nematodes [32]. Other studies attributed the antifungal activity of CAPE to its action on RNA, DNA, and cellular proteins, which are probable targets of this compound [13]. On the other hand, Su and coworkers suggested that the cytotoxicity of CAPE could be related to its apoptotic effect on the cells [33]. Marin and coworkers found that CAPE was able to induce the genes that are responsible for apoptosis and oxidative stress response. They reported that some polyphenolic compounds may have pro-oxidant activity, which can induce oxidative stress in the cells through the production of high levels of reactive oxygen species or inhibition of the system antioxidants [34]. In our study, we found that CAPE can induce apoptosis in five *Candida* strains, which are *C. albicans* ATCC 44829, *C. albicans* SZMC 1423, *C. albicans* SZMC 1424, *C. parapsilosis* SZMC 8007, *C. tropicalis* SZMC 1366, and *C. tropicalis* SZMC 1512. Surprisingly, *C. parapsilosis* SZMC 8007, *C. glabrata* SZMC 1374, and *C. glabrata* SZMC 1378 did not exhibit apoptotic cell death, which indicates that different species of *Candida* and, in one case, different strains of the same species have different cell death responses to CAPE. Moreover, the TEM images of CAPE-treated *Candida* cells showed the typical hallmarks of apoptosis in most *Candida* species. Notably, the apoptotic hallmarks were almost the same in different *Candida* species. Similar apoptotic hallmarks have been reported by several previous studies on *C. albicans*. De Nollin and Borgers (1975) reported the alterations of the surface micromorphology in *C. albicans* after treatment with miconazole. Shrinkage of protoplasm, abnormal cell and nuclear morphology, and vacuolization were also observed in plantaricin peptide-treated cells of *C. albicans* [35]. Distortion of the cell walls and membranes, which caused alterations of the surface micromorphology, could be explained due to a change in the permeability of the cell membrane, which could cause an osmotic imbalance, leading to alterations and indentations of the cell wall in collapsed cells [36].

Furthermore, we investigated the mechanisms involved in CAPE-induced apoptosis in *Candida* spp. Application of the broad-range pan-caspase inhibitor Z-VAD-FMK significantly reduced CAPE-induced apoptosis in *C. albicans* ATCC 44829, *C. albicans* SZMC 1424, *C. tropicalis* SZMC 1366, and *C. tropicalis* SZMC 1512. Such results suggest that this compound induced yeast caspase (Yca1p)-dependent apoptosis in these strains. Since CAPE can increase the permeability of the plasma membrane to ions [20], it can cause depolarization in mitochondria. This could lead to the release of cytochrome c and other proapoptotic factors into the cytosol, which in turn leads to the activation of yeast metacaspase Yca1p, resulting in the activation of caspase cascade inducing apoptosis [37]. However, the pre-incubation of CAPE-treated *C. albicans* SZMC 1423 and *C. parapsilosis* SZMC 8007 with the pan-caspase inhibitor Z-VAD-FMK did not affect their viability, which means that the CAPE-induced apoptosis in these strains was Yca1p-independent. This suggests that it could be due to the release of the apoptosis-inducing factor Aif1p from the mitochondria triggered by CAPE. These results support the fact that CAPE not only has species- and strain-dependent cell death responses in *Candida* but also could induce apoptotic cell death through different mechanisms.

4. Materials and Methods

4.1. Materials

For our experiments, CAPE (Sigma-Aldrich, Buchs, Switzerland); sodium dodecyl sulfate; crystal violet; peptone; yeast extract (Merck, Germany); agar-agar (Fluka, Buchs, Switzerland); a modified version of RPMI 1640 medium (containing dextrose 1.8% (*w*/*v*), MOPS 3.4% (*w*/*v*), and adenine 0.002% (*w*/*v*)) (Sigma-Aldrich, St. Louis, MI, USA); potassium dihydrogen phosphate; disodium hydrogen phosphate (Reanal, Budapest, Hungary); dimethyl sulfoxide; ethanol (VWR Chemicals, Paris, France); sodium chloride (VWR Chemicals, Debrecen, Hungary); glucose (VWR Chemicals, Leuven, Belgium); adenine; calcium chloride; magnesium chloride; potassium chloride (Scharlau Chemie S.A, Sentmenat, Spain); Z-VAD-FMK (Biovision, Milpitas, CA, USA); glutardialdehyde solution; osmium tetroxide; propylene oxide; Durcupan (R) ACM components A/M, B, C, and D (Sigma-Aldrich, Darmstadt, Germany); 0.22 µm vacuum filters (Merck Millipore, Guyancourt, France); sterile 96-well microtiter plates for susceptibility testing (Costar®, Phoenix, AZ, USA) and for biofilm assays (Sarstedt AG & Co. KG, Nümbrecht, Germany, Catalog number: 83.3924.500); CF®488A Annexin V and PI apoptosis Kit (Biotium, Fremont, CA, USA); and methanol (Chemolab Ltd., Budapest, Hungary) were used. All of the chemicals used in the experiments were of analytical or spectroscopic grade.

4.2. Instruments Used in the Experiments

A Thermo Scientific Heraeus B12 incubator (Thermo Fisher Scientific, Waltham, MA, USA), Sanyo orbital incubator (Sanyo, Japan), Sanyo autoclave (Sanyo, Japan), Hitachi U-2910 UV/Vis spectrophotometer (Hitachi, Japan), WTW pH meter (inoLab, Germany), Multiskan EX plate reader (Thermo Fisher Scientific Inc., Vantaa, Finland), benchtop centrifuge (Hettich, Buford, GA, USA), Ultramicrotome Reichert Jung Ultracut E (LabX, Midland, ON, Canada), JEOL-1200EX Transmission electron microscope (TEM), and Attune NxT flow cytometer (Thermo Fisher Scientific Inc., Massachusetts, USA) were used throughout the experiments.

A Thermo Scientific Heraeus B12 incubator (Thermo Fisher Scientific, Waltham, MA, USA), Sanyo orbital incubator (Sanyo, Japan), Sanyo autoclave (Sanyo, Japan), Hitachi U-2910 UV/Vis spectrophotometer (Hitachi, Japan), WTW pH meter (inoLab, Germany), Multiskan EX plate reader (Thermo Fisher Scientific Inc., Vantaa, Finland), benchtop centrifuge (Hettich, USA), Ultramicrotome Reichert Jung Ultracut E (LabX, Canada), JEOL-1200EX Transmission electron microscope (TEM), and Attune NxT flow cytometer (Thermo Fisher Scientific Inc., Massachusetts, USA) were used throughout the experiments.

4.3. Test Microorganisms, Culture Media, and Growth Conditions

Four species of *Candida* were used: *C. albicans*, *C. tropicalis*, *C. glabrata*, and *C. parapsilosis*. Nine strains were included, one of which was the American Type Culture Collection (ATCC) strain, and the others were *Candida* isolates obtained from Szeged Microbial Collection (SZMC), University of Szeged, Hungary (Table 2). All strains were maintained at the Department of General and Environmental Microbiology, Institute of Biology, University of Pécs, Hungary. All strains were grown in yeast extract peptone dextrose (YPD) broth (yeast extract 1%, peptone 2%, and glucose 2% in distilled water, pH 6.8) or on YPD plates.

Table 2. *Candida* species and strains used in the study.

Species	Collection Code	Origin	Biofilm-Forming Ability
C. albicans	ATCC 44829	auxotrophic mutant isolated after N-methyl N′-nitro-N-nitrosoguanidine treatment of a wild-type strain of *C. albicans*.	Non-biofilm forming
C. albicans	SZMC 1423	clinical sample/Debrecen, Hungary	Non-biofilm forming
C. albicans	SZMC 1424	clinical sample/Debrecen, Hungary	High-biofilm forming
C. tropicalis	SZMC 1366	hemoculture/Debrecen, Hungary	High-biofilm forming
C. tropicalis	SZMC 1512	-/Pécs, Hungary	Non-biofilm forming
C. glabrata	SZMC 1374	clinical sample/Debrecen, Hungary	High-biofilm forming
C. glabrata	SZMC 1378	clinical sample/Debrecen, Hungary	Non-biofilm forming
C. parapsilosis	SZMC 8007	clinical sample/Szeged, Hungary	High-biofilm forming
C. parapsilosis	SZMC 8008	unknown	Non-biofilm forming

4.4. *Preparation of Stock Solution of CAPE*

The stock solution was freshly prepared by dissolving CAPE in ethanol at a concentration of 10 mg/mL. The stock solution was kept in the freezer at −20 °C.

4.5. *Antifungal Susceptibility Testing*

The broth microdilution method was performed to determine the minimal inhibitory concentration (MIC_{80}) according to the protocol of the National Committee for Clinical Laboratory Standards Institute with some modifications [38]. Briefly, a standardized initial inoculum (10^6 cells/mL) was applied in all experiments. The experiments were performed in sterile, flat-bottom 96-well microplates. To obtain final CAPE concentrations ranging from 400 to 3.125 µg/mL, equal volumes (100 µL) of cell suspension and CAPE-containing YPD medium were added into the wells. Negative controls (media and cell suspension without CAPE) and blanks (media with CAPE) were included in each experiment. The concentration of the solvent was constantly fixed at 1%. The plates were incubated at 35 °C, and the absorbance was measured after 48 h at 600 nm using a Multiskan EX plate reader. The MIC_{80} of CAPE was determined as the lowest concentration that causes an 80% reduction in the growth when compared with that of the negative control.

4.6. *Biofilm-Forming Ability Assay*

Biofilm formation assay was performed using the crystal violet staining method as described previously [39]. To inoculate the test microplates, a stationary-phase yeast culture was prepared using an inoculum size equivalent to 0.5 McFarland standard. The culture was shaken thoroughly and then diluted at 1:100 using RPMI-1640 medium. A series of two-fold dilutions were prepared from the stock solution of CAPE. In the test microplates, equal volumes (100 µL) of each dilution were added to identical volumes (100 µL) of the diluted cell suspensions to obtain final concentrations from 100 to 1.562 µg/mL CAPE in the wells. In each experiment, the negative controls and blanks were included. The concentration of the solvent was constantly fixed at 1%. The microplates were kept in an incubator at 35 °C for 48 h; afterward, the liquid part was discarded and the remaining biofilms were repeatedly washed with phosphate-buffered saline (PBS) (pH 7.4). Formalin in PBS 2% (v/v) was used to fix the biofilms; then, the crystal violet 0.13% (w/v) staining was applied for 20 min at room temperature. The excess crystal violet stain was discarded, and the wells were washed thoroughly and repeatedly with PBS buffer. The estimation of biofilm mass was conducted by adding sodium dodecyl sulfate (SDS) in ethanol (1% w/v) solution to each well to extract the stain overnight, and the absorbance of the solution was measured at 600 nm using a Multiskan EX plate reader. The minimum biofilm inhibitory concentration (MBIC) was defined as the lowest concentration of CAPE that was able to inhibit 90% of the biofilm-forming ability.

4.7. Biofilm Eradication Assay

The effect of CAPE on mature biofilms was verified as described by Nostro and coworkers [40]. Briefly, for the inoculation of the assay microplates, stationary-phase yeast cultures were prepared using an inoculum size equivalent to 0.5 McFarland standard and thereafter diluted 1:100 using RPMI-1640 medium. Microplates containing diluted cell suspension were kept in an incubator at 35 °C for 48 h. After the biofilm maturation, CAPE treatment was applied. Accordingly, the original RPMI culture was discarded and replaced with a CAPE-containing RPMI medium with concentrations ranging from 100 to 1.562 μg/mL. Negative controls and blanks were included in each experiment. The concentration of the solvent was constantly fixed at 1%. After 48 h of incubation at 35 °C, the liquid part of the media was discarded, and the remaining biofilms were washed, fixed, stained, and estimated as mentioned in the previous section (Section 4.6).

4.8. Biosorption of CAPE by Candida Cells

To determine the cellular biosorption of CAPE, YPD broth cultures of *Candida* strains were grown overnight at 35 °C and 150 rpm in an orbital shaker. The number of yeast cells was adjusted to 10^7 cells/mL in each case, and the cultures were treated with 100 μg/mL CAPE and incubated at 35°C with shaking at 150 rpm. The concentration of the solvent was constantly fixed at 1%. Samples were taken at the time points 0, 5, 10, 15, 20, 30, 60, and 120 min after admission and centrifuged (5000 rpm, 5 min), and the absorbance of the cell-free supernatants was measured at 330 nm (absorption maximum of CAPE) using a Hitachi U-2910 UV/Vis spectrophotometer. A calibration curve of two-fold serial dilutions of CAPE from 100 to 0.781 μg/mL was constructed and used to evaluate the biosorption levels of *Candida* cells [41].

4.9. Cell Death Examination Assay

The YPD broth media were inoculated with *Candida* cells (10^6 cell/mL) from fresh YPD plate cultures and incubated at 35 °C with shaking at 150 rpm until reaching the mid-exponential phase depending on their growth curves. Media containing sub-lethal concentrations (MIC_{80}) of CAPE were inoculated with 2.5×10^6 cells/mL of the mid-exponential phase cultures of different *Candida* species and strains and incubated at 35 °C with shaking for 3 h. Untreated cell samples were included as negative controls in each experiment. The concentration of the solvent was constantly fixed at 1% in all experiments. After the incubation period, cells were harvested and washed with PBS. The CF®488A Annexin V and PI apoptosis Kit was used according to the manufacturer's instructions to identify apoptosis and necrosis. In brief, *Candida* cells were re-suspended in 1X annexin V binding buffer at a concentration of 5×10^6 cells/mL. To 100 μL of this solution, 5 μL of CF®488A Annexin V and 2 μL of PI working solution were added. The tubes were gently vortexed and incubated for 20 minutes at room temperature in the dark. After incubation, 400 μL of 1X annexin V binding buffer was added to each tube and analyzed using an Attune NxT flow cytometer. Annexin V is responsible for the detection of phosphatidylserine translocation from the inner to outer leaflets of the plasma membrane, whereas PI is a membrane-impermeant DNA-binding dye that is usually used to selectively stain dead cells in a cell population. PI is excluded by living cells and early apoptotic cells but stains necrotic and late apoptotic cells with compromised membrane integrity.

4.10. Caspase Inhibitor Assay

Caspase inhibitor assay was performed as described by Yue and coworkers [42] with some modifications. Briefly, cells were divided into two groups. The first group was pretreated for 1 h at 35 °C with the broad-spectrum caspase inhibitor Z-VAD-FMK (final concentration 77 μM) before incubation with CAPE. The second group was used as a control (not treated with the caspase inhibitor). For microplate assays, the cells were harvested by centrifugation, washed twice with PBS, and then re-suspended in YPD broth. The cell density was adjusted to 2×10^6 cell/mL. Equal volumes (100 μL) of cell suspension and CAPE-containing YPD medium were dispensed into the wells to obtain the final CAPE concentration equal to the MIC_{80} of each strain. Negative controls (media and cell suspension without CAPE) and blanks (media with CAPE) were included in each experiment. The concentration of the solvent was constantly fixed at 1%. The plates were incubated at 35 °C, and the absorbance was measured after 48 h at 600 nm using a Multiskan EX plate reader.

4.11. Ultrastructural Examination of Candida Species by TEM

Media containing sub-lethal concentrations of CAPE were inoculated with 2.5×10^6 cells/mL of the mid-exponential phase cultures of different species of *Candida* and incubated at 35 °C with shaking for 3 h to induce apoptosis. After the incubation period, the cells were harvested by centrifugation (5000 rpm, 5 min). The pellets were immediately washed and re-suspended with modified PBS (a mixture of 50 mM K_2HPO_4 and KH_2PO_4 (pH 7.0), supplemented with 0.5 mM $MgCl_2$) and incubated at room temperature for 15 min to achieve equilibrium. Then, the samples were fixed overnight in 2.5% glutaraldehyde fixative buffered with modified PBS. The samples were then washed 4 times with modified PBS, and after short centrifugal sedimentation (1000 rpm, 2 min) preparation continued with a 2 % osmium tetroxide post-fixation on ice for 2 h. The cells were then washed twice with distilled water for 15 min and stained 'en bloc' in 1% aqueous uranyl acetate for 30 min. After two further washing steps with distilled water and short sedimentation, the cells were dehydrated in 70, 96, and 100% ethanol for 15 min each, subsequently. The cells were treated with propylene oxide twice for 10 min each time and then infiltrated for 1 h in a propylene oxide/Durcupan epoxy resin mixture (1:1) at room temperature. After 1 h, the cells were transferred to fresh epoxy resin drops for another 1 h. The resin was then changed, and the samples were left overnight in the fresh resin drops at room temperature. On the next day, the resin was changed twice while incubating at 40 °C for 2 h, subsequently. Finally, the samples were encapsulated in fresh epoxy resin and left at 56 °C for a two-day-long polymerization. Serial ultrathin sections were cut with Reichert Ultramicrotome, collected onto 300 mesh Nickel grids, counterstained on drops of uranyl acetate and Reynolds solution of lead citrate, washed thoroughly in sterile distilled water, and examined with a JEOL-1200 EX TEM at 80 KeV [43].

4.12. Statistical Analysis

All assays were carried out in triplicate, and data were expressed as mean ± standard deviation (SD). For data processing and visualization of the results, Microsoft Office Excel 2016 was used. Data were statistically analyzed through two-sample t-tests using Past3.21 software (University in Oslo, Oslo, Norway).

5. Conclusions

CAPE could be considered a promising natural antifungal agent. It has a concentration- and strain-dependent effect on the viability and biofilm-forming ability of the different *Candida* species. Moreover, it has a partial ability to eradicate the mature biofilms of biofilm-forming strains of *Candida*. In most *Candida* species and strains, the antifungal mechanism involves the induction of apoptotic cell death in treated cells. However, in other *Candida* species and strains, no apoptotic cell death was observed. This information suggests that CAPE may have a species- and strain-dependent cell death response in *Candida*.

Supplementary Materials: The following are available online at https://www.mdpi.com/article/10.3390/antibiotics10111359/s1, Figure S1: Representative flow cytometry scatter plot showing apoptosis of *C. albicans* ATCC 44829 treated with 50 μg/mL of CAPE, Figure S2: Representative flow cytometry scatter plot showing apoptosis of *C. albicans* SZMC 1423 treated with 100 μg/mL of CAPE, Figure S3: Representative flow cytometry scatter plot showing apoptosis of *C. albicans* SZMC 1424 treated with 50 μg/mL of CAPE, Figure S4: Representative flow cytometry scatter plot showing apoptosis of *C. parapsilosis* SZMC 8007 treated with 25 μg/mL of CAPE, Figure S5: Representative flow cytometry scatter plot showing apoptosis of *C. parapsilosis* SZMC 8008 treated with 12.5 μg/mL of CAPE, Figure S6: Representative flow cytometry scatter plot showing apoptosis of *C. glabrata* SZMC 1374 treated with 12.5 µg/mL of CAPE, Figure S7: Representative flow cytometry scatter plot showing apoptosis of C. glabrata SZMC 1378 treated with 12.5 µg/mL of CAPE.

Author Contributions: Conceptualization, I.A. and G.P.; methodology, I.A., A.V. and E.P.; validation, I.A, E.P. and G.P.; formal analysis, all authors; investigation, I.A. and E.P.; resources, I.A., E.P., A.V. and G.P.; data curation, I.A.; writing—original draft preparation, I.A. and G.P.; writing—review and editing, I.A., E.P., Á.C., A.V., S.D. and G.P.; visualization, I.A., E.P., Á.C., A.V., S.D. and G.P.; supervision, G.P. and I.A.; project administration, G.P. All authors have read and agreed to the published version of the manuscript.

Funding: This research received no external funding.

Institutional Review Board Statement: Not applicable.

Informed Consent Statement: Not applicable.

Data Availability Statement: Not applicable.

Acknowledgments: This research was also connected to the project GINOP-2.3.3-15-2016-00006 (Széchenyi 2020 Programme).

Conflicts of Interest: The authors declare no conflict of interest.

Abbreviations

CAPE: Caffeic Acid Phenethyl Ester; MIC, Minimal Inhibitory Concentration; MBIC, Minimal Biofilm Inhibitory Concentration; PBS, Phosphate-Buffered Saline; PI, Propidium Iodide; YPD, Yeast extract Peptone Dextrose.

References

1. Muñoz, J.E.; Rossi, D.C.P.; Jabes, D.L.; Barbosa, D.A.; Cunha, F.F.M.; Nunes, L.R.; Arruda, D.C.; Pelleschi Taborda, C. *In Vitro* and *In Vivo* Inhibitory Activity of Limonene against Different Isolates of *Candida* spp. *J. Fungi* **2020**, *6*, 183. [CrossRef]
2. Walsh, T.J.; Dixon, D.M. Spectrum of mycoses: Chapter 75. In *Medical Microbiology*, 4th ed.; Baron, S., Ed.; University of Texas Medical Branch at Galveston: Galveston, TX, USA, 1996.
3. Patil, S.; Rao, R.S.; Majumdar, B.; Anil, S. Clinical appearance of oral *Candida* infection and therapeutic strategies. *Front. Microbiol.* **2015**, *6*, 1391. [CrossRef] [PubMed]
4. Silva, S.; Henriques, M.; Martins, A.; Oliveira, R.; Williams, D.; Azeredo, J. Biofilms of non-*Candida albicans Candida* species: Quantification, structure and matrix composition. *Sabouraudia* **2009**, *47*, 681–689. [CrossRef]
5. Ghannoum, M.A. *Candida*: A causative agent of an emerging infection. *J. Investig. Dermatol. Symp. Proc.* **2001**, *6*, 188–196. [CrossRef] [PubMed]
6. Donadu, M.G.; Peralta-Ruiz, Y.; Usai, D.; Maggio, F.; Molina-Hernandez, J.B.; Rizzo, D.; Bussu, F.; Rubino, S.; Zanetti, S.; Paparella, A. Colombian Essential Oil of Ruta graveolens against Nosocomial Antifungal Resistant *Candida* Strains. *J. Fungi* **2021**, *7*, 383. [CrossRef]
7. Tsui, C.; Kong, E.F.; Jabra-Rizk, M.A. Pathogenesis of *Candida albicans* biofilm. *FEMS Pathog. Dis.* **2016**, *74*, 1–13. [CrossRef]
8. El-Houssaini, H.H.; Elnabawy, O.M.; Nasser, H.A.; Elkhatib, W.F. Correlation between antifungal resistance and virulence factors in *Candida albicans* recovered from vaginal specimens. *Microb. Pathog.* **2019**, *128*, 13–19. [CrossRef]
9. Soliman, S.; Alnajdy, D.; El-Keblawy, A.A.; Mosa, K.A.; Khoder, G.; Noreddin, A.M. Plants' natural products as alternative promising anti-*Candida* drugs. *Pharmacogn. Rev.* **2017**, *11*, 104. [CrossRef]
10. Mukherjee, P.K.; Chandra, J. *Candida* biofilm resistance. *Drug Resist. Updatates* **2004**, *7*, 301–309. [CrossRef] [PubMed]
11. Seneviratne, C.J.; Jin, L.; Samaranayake, L.P. Biofilm lifestyle of *Candida*: A mini review. *Oral Dis.* **2008**, *14*, 582–590. [CrossRef]
12. Alfarrayeh, I.I.S. Bioactivities and Potential Beneficial Properties of Propolis Ethanolic Extract, Caffeic acid phenethyl Ester, and Arabic Coffee Beans Extract. Ph.D. Thesis, University of Pécs, Pécs, Hungary, 2021.
13. Murtaza, G.; Karim, S.; Akram, M.R.; Khan, S.A.; Azhar, S.; Mumtaz, A.; Bin Asad, M.H.H. Caffeic acid phenethyl ester and therapeutic potentials. *Biomed Res. Int.* **2014**, *2014*, 145342. [CrossRef]
14. Koru, Ö.; Avcu, F.; Tanyüksel, M.; Ural, A.U.; Araz, R.E.; Şener, K. Cytotoxic effects of caffeic acid phenethyl ester (CAPE) on the human multiple myeloma cell line. *Turkish J. Med. Sci.* **2009**, *39*, 863–870. [CrossRef]
15. Huang, S.; Zhang, C.P.; Wang, K.; Li, G.Q.; Hu, F.L. Recent advances in the chemical composition of propolis. *Molecules* **2014**, *19*, 19610–19632. [CrossRef]
16. Yordanov, Y. Caffeic acid phenethyl ester (CAPE): Pharmacodynamics and potential for therapeutic application. *Pharmacia* **2019**, *66*, 107. [CrossRef]

17. Alfarrayeh, I.; Fekete, C.; Gazdag, Z.; Papp, G. Propolis ethanolic extract has double-face in vitro effect on the planktonic growth and biofilm formation of some commercial probiotics. *Saudi J. Biol. Sci.* **2021**, *28*, 1033–1039. [CrossRef] [PubMed]
18. Cui, K.; Lu, W.; Zhu, L.; Shen, X.; Huang, J. Caffeic acid phenethyl ester (CAPE), an active component of propolis, inhibits Helicobacter pylori peptide deformylase activity. *Biochem. Biophys. Res. Commun.* **2013**, *435*, 289–294. [CrossRef]
19. Meyuhas, S.; Assali, M.; Huleihil, M.; Huleihel, M. Antimicrobial activities of caffeic acid phenethyl ester. *J. Mol. Biochem.* **2015**, *4*.
20. Mirzoeva, O.K.; Grishanin, R.N.; Calder, P.C. Antimicrobial action of propolis and some of its components: The effects on growth, membrane potential and motility of bacteria. *Microbiol. Res.* **1997**, *152*, 239–246. [CrossRef]
21. Lv, L.; Cui, H.; Ma, Z.; Liu, X.; Yang, L. Recent progresses in the pharmacological activities of caffeic acid phenethyl ester. *Naunyn Schmiedebergs Arch. Pharmacol.* **2021**, *394*, 1327–1339. [CrossRef]
22. Niu, Y.; Wang, K.; Zheng, S.; Wang, Y.; Ren, Q.; Li, H.; Ding, L.; Li, W.; Zhang, L. Antibacterial effect of caffeic acid phenethyl ester on cariogenic bacteria and Streptococcus mutans biofilms. *Antimicrob. Agents Chemother.* **2020**, *64*, e00251-20. [CrossRef] [PubMed]
23. De Barros, P.P.; Rossoni, R.D.; Garcia, M.T.; de Lima Kaminski, V.; Loures, F.V.; Fuchs, B.B.; Mylonakis, E.; Junqueira, J.C. The anti-biofilm efficacy of caffeic acid phenethyl ester (CAPE) in vitro and a murine model of oral candidiasis. *Front. Cell. Infect. Microbiol.* **2021**, *11*, 700305. [CrossRef]
24. Sun, L.; Liao, K.; Hang, C. Caffeic acid phenethyl ester synergistically enhances the antifungal activity of fluconazole against resistant *Candida albicans*. *Phytomedicine* **2018**, *40*, 55–58. [CrossRef]
25. Sun, L.; Hang, C.; Liao, K. Synergistic effect of caffeic acid phenethyl ester with caspofungin against *Candida albicans* is mediated by disrupting iron homeostasis. *Food Chem. Toxicol.* **2018**, *116*, 51–58. [CrossRef]
26. Michalak, I.; Chojnacka, K.; Witek-Krowiak, A. State of the art for the biosorption process—A review. *Appl. Biochem. Biotechnol.* **2013**, *170*, 1389–1416. [CrossRef]
27. Sutcu, M.; Salman, N.; Akturk, H.; Dalgıc, N.; Turel, O.; Kuzdan, C.; Kadayifci, E.K.; Sener, D.; Karbuz, A.; Erturan, Z. Epidemiologic and microbiologic evaluation of nosocomial infections associated with *Candida* spp in children: A multicenter study from Istanbul, Turkey. *Am. J. Infect. Control* **2016**, *44*, 1139–1143. [CrossRef] [PubMed]
28. Bona, E.; Cantamessa, S.; Pavan, M.; Novello, G.; Massa, N.; Rocchetti, A.; Berta, G.; Gamalero, E. Sensitivity of *Candida albicans* to essential oils: Are they an alternative to antifungal agents? *J. Appl. Microbiol.* **2016**, *121*, 1530–1545. [CrossRef] [PubMed]
29. Mandras, N.; Roana, J.; Scalas, D.; Del Re, S.; Cavallo, L.; Ghisetti, V.; Tullio, V. The Inhibition of Non-*albicans Candida* Species and Uncommon Yeast Pathogens by Selected Essential Oils and Their Major Compounds. *Molecules* **2021**, *26*, 4937. [CrossRef]
30. Possamai Rossatto, F.C.; Tharmalingam, N.; Escobar, I.E.; D'Azevedo, P.A.; Zimmer, K.R.; Mylonakis, E. Antifungal Activity of the Phenolic Compounds Ellagic Acid (EA) and Caffeic Acid Phenethyl Ester (CAPE) against Drug-Resistant *Candida* auris. *J. Fungi* **2021**, *7*, 763. [CrossRef] [PubMed]
31. Cigut, T.; Polak, T.; Gasperlin, L.; Raspor, P.; Jamnik, P. Antioxidative activity of propolis extract in yeast cells. *J. Agric. Food Chem.* **2011**, *59*, 11449–11455. [CrossRef]
32. Breger, J.; Fuchs, B.B.; Aperis, G.; Moy, T.I.; Ausubel, F.M.; Mylonakis, E. Antifungal chemical compounds identified using a *C. elegans* pathogenicity assay. *PLoS Pathog.* **2007**, *3*, e18. [CrossRef]
33. Su, Z.Z.; Lin, J.; Prewett, M.; Goldstein, N.I.; Fisher, P.B. Apoptosis mediates the selective toxicity of caffeic acid phenethyl ester (CAPE) toward oncogene-transformed rat embryo fibroblast cells. *Anticancer Res.* **1995**, *15*, 1841–1848.
34. Marin, E.H.; Paek, H.; Li, M.; Ban, Y.; Karaga, M.K.; Shashidharamurthy, R.; Wang, X. Caffeic acid phenethyl ester exerts apoptotic and oxidative stress on human multiple myeloma cells. *Investig. New Drugs* **2019**, *37*, 837–848. [CrossRef] [PubMed]
35. Sharma, A.; Srivastava, S. Anti-*Candida* activity of two-peptide bacteriocins, plantaricins (Pln E/F and J/K) and their mode of action. *Fungal Biol.* **2014**, *118*, 264–275. [CrossRef] [PubMed]
36. De Nollin, S.; Borgers, M. Scanning electron microscopy of *Candida albicans* after in vitro treatment with miconazole. *Antimicrob. Agents Chemother.* **1975**, *7*, 704–711. [CrossRef] [PubMed]
37. Lone, S.A.; Wani, M.Y.; Fru, P.; Ahmad, A. Cellular apoptosis and necrosis as therapeutic targets for novel Eugenol Tosylate congeners against *Candida albicans*. *Sci. Rep.* **2020**, *10*, 1–15. [CrossRef]
38. CLSI. *Reference Method for Broth Dilution Antifungal Susceptibility Testing of Yeasts; Approved Standard*, 2nd ed.; Clinical and Laboratory Standards Institute: Pittsburgh, PA, USA, 2002; Volume 22, ISBN 1562384694.
39. Stepanović, S.; Vuković, D.; Hola, V.; Di Bonaventura, G.; Djukić, S.; Ćirković, I.; Ruzicka, F. Quantification of biofilm in microtiter plates: Overview of testing conditions and practical recommendations for assessment of biofilm production by staphylococci. *Apmis* **2007**, *115*, 891–899. [CrossRef] [PubMed]
40. Nostro, A.; Roccaro, A.S.; Bisignano, G.; Marino, A.; Cannatelli, M.A.; Pizzimenti, F.C.; Cioni, P.L.; Procopio, F.; Blanco, A.R. Effects of oregano, carvacrol and thymol on Staphylococcus aureus and Staphylococcus epidermidis biofilms. *J. Med. Microbiol.* **2007**, *56*, 519–523. [CrossRef]
41. Sun, X.; Zhao, Y.; Liu, L.; Jia, B.; Zhao, F.; Huang, W.; Zhan, J. Copper tolerance and biosorption of Saccharomyces cerevisiae during alcoholic fermentation. *PLoS ONE* **2015**, *10*, e0128611. [CrossRef] [PubMed]
42. Yue, Q.; Zhou, X.; Leng, Q.; Zhang, L.; Cheng, B.; Zhang, X. 7-ketocholesterol-induced caspase-mediated apoptosis in Saccharomyces cerevisiae. *FEMS Yeast Res.* **2013**, *13*, 796–803. [CrossRef] [PubMed]
43. Phillips, A.J.; Sudbery, I.; Ramsdale, M. Apoptosis induced by environmental stresses and amphotericin B in *Candida albicans*. *Proc. Natl. Acad. Sci. USA* **2003**, *100*, 14327–14332. [CrossRef]

 antibiotics

Article

Can Vitamin B12 Assist the Internalization of Antisense LNA Oligonucleotides into Bacteria?

Sara Pereira [1], Ruwei Yao [2], Mariana Gomes [1], Per Trolle Jørgensen [2], Jesper Wengel [2], Nuno Filipe Azevedo [1] and Rita Sobral Santos [1,*]

1 Laboratory for Process Engineering, Environment, Biotechnology and Energy (LEPABE), Faculty of Engineering, University of Porto, R. Dr. Roberto Frias, 4200-465 Porto, Portugal; up201610825@fe.up.pt (S.P.); mggomes@fe.up.pt (M.G.); nazevedo@fe.up.pt (N.F.A.)

2 Biomolecular Nanoscale Engineering Center, Department of Physics, Chemistry and Pharmacy, University of Southern Denmark, Campusvej 55, 5230 Odense M, Denmark; ruweiy@kemi.dtu.dk (R.Y.); ptj@sdu.dk (P.T.J.); jwe@sdu.dk (J.W.)

* Correspondence: ritasantos@fe.up.pt

Abstract: The emergence of bacterial resistance to traditional small-molecule antibiotics is fueling the search for innovative strategies to treat infections. Inhibiting the expression of essential bacterial genes using antisense oligonucleotides (ASOs), particularly composed of nucleic acid mimics (NAMs), has emerged as a promising strategy. However, their efficiency depends on their association with vectors that can translocate the bacterial envelope. Vitamin B_{12} is among the largest molecules known to be taken up by bacteria and has very recently started to gain interest as a trojan-horse vector. Gapmers and steric blockers were evaluated as ASOs against *Escherichia coli* (*E. coli*). Both ASOs were successfully conjugated to B_{12} by copper-free azide-alkyne click-chemistry. The biological effect of the two conjugates was evaluated together with their intracellular localization in *E. coli*. Although not only B_{12} but also both B_{12}-ASO conjugates interacted strongly with *E. coli*, they were mostly colocalized with the outer membrane. Only 6–9% were detected in the cytosol, which showed to be insufficient for bacterial growth inhibition. These results suggest that the internalization of B_{12}-ASO conjugates is strongly affected by the low uptake rate of the B_{12} in *E. coli* and that further studies are needed before considering this strategy against biofilms in vivo.

Keywords: antibacterial drug; vitamin B_{12}; antisense oligonucleotides; nucleic acid mimics; LNA; 2′OMe

Citation: Pereira, S.; Yao, R.; Gomes, M.; Jørgensen, P.T.; Wengel, J.; Azevedo, N.F.; Sobral Santos, R. Can Vitamin B12 Assist the Internalization of Antisense LNA Oligonucleotides into Bacteria? *Antibiotics* **2021**, *10*, 379. https://doi.org/10.3390/antibiotics10040379

Academic Editors: Nicholas Dixon and Marc Maresca

Received: 17 February 2021
Accepted: 1 April 2021
Published: 3 April 2021

1. Introduction

The emergence of bacterial resistance to traditional antibiotics is considered a major threat in modern medicine [1,2]. Inevitably, innovative research focused on different antibacterial strategies is needed. Antisense oligonucleotides (ASOs) designed to inhibit bacterial gene expression have been gaining increased interest in recent years [3]. ASOs are especially interesting because even if bacteria develop a mutation that renders them resistant (one of the most common forms of resistance), the ASO can be easily redesigned to become an effective antibacterial drug again [4]. ASOs composed of nucleic acid mimics (NAMs), and in particular, locked nucleic acids (LNAs), possess improved target specificity, binding affinity, and resistance to exo- and endonucleases, compared to unmodified RNA or DNA [5,6], and have been successfully tested for clinical applications [7,8]. ASOs can be divided into two major categories, according to their mechanism of action: RNase H competent (or gapmers) and steric blockers (Figure 1). Gapmers are composed of DNA monomers that are typically flanked by LNA or other RNA-mimicking monomers. Upon hybridization of the gapmer to the target mRNA, the RNase H enzyme recognizes the DNA-mRNA heteroduplex and cleaves the target mRNA, leading to its degradation. Alternatively, the hybridization of steric blockers to the target mRNA simply physically blocks the access of the RNA polymerase to the target mRNA, thus directly inhibiting its translation [9–11]. There is only one study reporting the use of gapmers to target bacteria [11].

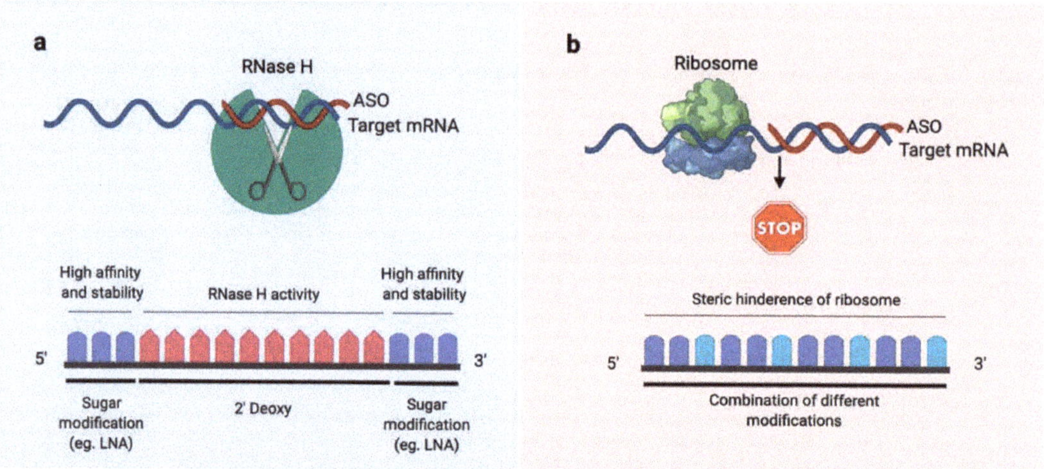

Figure 1. Different mechanisms that play a role in the modulation of the RNA function in bacteria. (**a**) Upon hybridization of a gapmer (red), the RNase H is recruited, and the target is degraded. (**b**) Steric hindrance of the ribosome caused by the hybridization between a steric blocker (red) and the complementary mRNA sequence. Figure created using BioRender.

Nonetheless, the use of ASOs is limited by their inability to efficiently penetrate the complex envelope of bacteria. To overcome this burden, delivery vectors mostly focused on cell-penetrating peptides (CPPs) have been investigated. However, CPPs have been almost exclusively conjugated to neutral NAMs, such as peptide nucleic acids (PNAs) or phosphorodiamidate morpholino oligos (PMOs), which might present cytotoxicity and solubility issues [12–14]. Due to chemical conjugation difficulties, the vectorization of anionic ASOs with CPPs has been hampered. In a different approach, vitamin B_{12} (B_{12} or cobalamin), one of the largest molecules known to be taken up by bacteria, can be considered as a trojan-horse vector for neutral as well as anionic ASOs. The uptake system of B_{12} has been mostly studied in *E. coli*. [15]. *E. coli* uptakes B_{12} through the outer-membrane β-barrel protein BtuB in a TonB-dependent manner [16]. In the periplasm, BtuF binds to and delivers B_{12} to the ABC-type transporter BtuCD in the inner membrane, which, in turn, transports B_{12} into the cytoplasm [17,18].

Several functional groups are available for the modification of B_{12} to allow conjugation with ASOs, but not all modification sites are suitable to sustain their recognition and uptake [19]. Chromiński et al. described for the first time the synthesis of a clickable B_{12} derivate, possessing an azide functionality at the 5′ end [20]. This modification has already been tested for the copper-dependent conjugation of B_{12} with oligonucleotides, either composed of PNA or 2′OMe, mainly to inhibit genes that code for reporter proteins such as the red fluorescent protein (RFP) [6,20,21]. To our knowledge, there is only one study where a B_{12} conjugate was studied to decrease bacterial growth by inhibition of the essential gene *acpP* in *E. coli* using a PNA ASO [22]. This B_{12}-PNA conjugate was only proved to inhibit *E. coli* growth in a very specific medium named Scarlett and Turner [22]. Under these conditions, even in the absence of B_{12} conjugates, bacteria only started growing after 48 h, while in other common minimal media, the exponential growth starts already after 5 h [23].

Occasionally, infections are associated with the formation of biofilms, adding an extra barrier for the use of antibacterial compounds [24]. ASOs were already shown to prevent biofilm formation and reduce mature biofilms, using either peptide nucleic acids (PNAs) or phosphorodiamidate morpholino oligomers (PMOs) as NAMs, conjugated to CPPs [25,26] but, to the best of our knowledge, never conjugated to B_{12}. However, and because the cytosol is the ultimate target for these conjugates, it is important to first investigate their single-cell internalization.

In this study, we have investigated, for the first time, the internalization and inhibition efficiency of two different copper-free clicked conjugates, composed of vitamin B_{12} linked to LNA-based ASOs: a gapmer and a steric blocker. Both ASOs were designed to target the *acpP* gene in *E. coli*, which codes for an essential protein involved in fatty acid biosynthesis [27].

2. Results and Discussion

2.1. Conjugation of ASOs with Vitamin B_{12}

In this study, two different kinds of LNA antisense oligonucleotides were designed and synthesized to target the *acpP* gene in *E. coli* (Figure 2a): an LNA/DNA gapmer (ASO_{gapmer}) and an LNA/2′OMe steric blocker (ASO_{steric}) [28,29]. While steric blockers have been widely tested in bacteria [4,27,30–32], gapmers have been mostly studied in mammalian cells [11]. As such, we intended to compare the potency of the different antisense mechanisms in *E. coli*. As the bacterial envelope poses a stringent barrier to the internalization of oligonucleotides, appropriate vectors must be applied for their transport into the bacterial cytosol. This is the first study documenting the use of B_{12} as a vector for LNA oligonucleotides. For the association of B_{12} to ASO_{steric} and ASO_{gamper}, a copper-free ring-strain-promoted azide-alkyne coupling reaction was used (Figure 2b) [33]. This method proved to be efficient and resulted in satisfactory yields (Table S1). The increase in HPLC retention time for the B_{12}-ASO_{gamper} and B_{12}-ASO_{steric} compared with the ASO_{gapmer} and ASO_{steric} points toward efficient conjugation, confirmed by the obtained molecular masses, which are similar to the calculated theoretical values (Figure S1). The conjugation yields were determined as 70% and 97%, respectively, for B_{12}-ASO_{gamper} and B_{12}-ASO_{steric}.

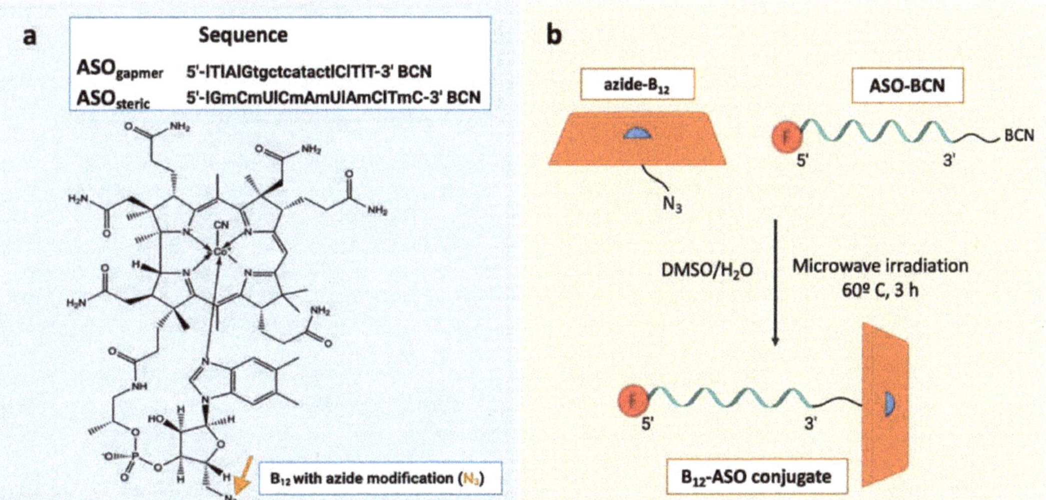

Figure 2. B_{12} and antisense oligonucleotides (ASOs) conjugation: sequences and structures. (**a**) Sequence of the ASO_{gapmer} and the ASO_{steric} (LNA nucleotide monomers are represented with upper case letters preceded by l, 2′OMe monomers are represented with upper case letters preceded by m, and DNA monomers are represented by lower case letters) and structure of 5′-end azide-modified B_{12} used in this study. The arrow points to the region of conjugation. (**b**) Schematic illustration of the synthesis of the B_{12}-ASO conjugates through copper-free azide-alkyne chemistry.

2.2. Bacterial Susceptibility Tests

Both ASOs were designed to recognize the *acpP* essential gene in *E. coli* and thus inhibit its expression. This should result in decreased *E. coli* viability, as long as the ASOs conjugated to B_{12} can efficiently penetrate the bacterial envelope. We investigated the ability of both conjugates (B_{12}-ASO_{gapmer} and B_{12}-ASO_{steric}) to inhibit the growth of *E. coli* in Davis minimal medium at a concentration of 30 μM. This concentration was selected based on the good inhibition efficiency of a cell-penetrating peptide (CPP) conjugated with PNA, targeting the same gene (Figure 3, orange line). Our results show no inhibition using either conjugate composed of B_{12} at the same concentration (Figure 3).

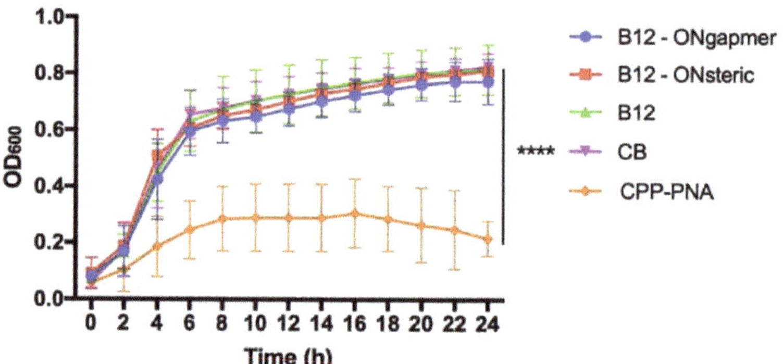

Figure 3. Growth of *E. coli* K12 in Davis minimal medium supplemented with B_{12}-ASO_{gapmer}, B_{12}-ASO_{steric}, and B_{12} (at a concentration of 30 μM). CB represents the bacterial growth control in medium without any supplementation. Growth inhibition of *E. coli* K12 using a cell-penetrating peptide conjugated with an ASO composed of peptide nucleic acids (PNAs) (cell-penetrating peptides (CPP)-PNA) at a concentration of 30 μM is also shown. Results from three independent experiments (using duplicates in each) are presented as mean values and respective standard deviations. Statistical differences are indicated when appropriate in * ($p \leq 0.0001$, ****).

In previous studies, PNA and 2′OMe steric blockers conjugated to B_{12} were able to decrease by 1-fold the expression of red fluorescence protein (RFP) in *E. coli* in Davis minimal medium [6,21]. However, to our knowledge, there is no other study including a regular growth control where B_{12}-ASOs were investigated to kill bacteria, targeting an essential gene rather than a report protein. The only existing study uses a B_{12}-PNA (ASO_{steric} targeting *acpP*) against *E. coli* in a particular medium where the control bacteria only starts growing after 48 h [22]. Nonetheless, the activity of this ASO sequence is already well established, as growth inhibition of *E. coli* K12 has been repeatedly reported using CPP-PNA [27,34,35], which was also confirmed herein. It is clear from the growth curves that the internalization occurs using the CPP as a vector for ASOs, as opposed to the B_{12} vector.

The lack of inhibitory effect of the *E. coli* K12 growth, observed with the conjugates synthesized in the present work, raises the question if the conjugates were efficiently internalized in the bacterial cells. In order to answer this question, location studies were performed next.

2.3. Evaluation of the Internalization of B_{12}-ASOs

To examine the internalization of both conjugates in *E. coli* K12 and assess if association of the ASOs to the B_{12} could have hampered B_{12}-promoted uptake, bacteria were observed under an epifluorescence microscope, after incubation with each of the Cy3-labeled conjugates or controls (B_{12}, ASO_{gapmer}, and ASO_{steric}).

As expected, almost no fluorescent bacteria were detected when ASO_{gapmer} and ASO_{steric} were used alone (Figure 4–ASO_{gapmer} and ASO_{steric}, Cy3 line). In contrast, it is clear that the conjugation of B_{12} to either ASO significantly increased the amount of fluorescently labeled *E. coli* K12, with all cells becoming fluorescent (Figure 4, B_{12}-ASO_{gapmer} and B_{12}-ASO_{steric}). The same was observed for the B_{12} control (Figure 4, B_{12}). Figure 4 shows images obtained at 30 μM, but a similar pattern was obtained for the lower concentration tested (15 μM, Figure S2).

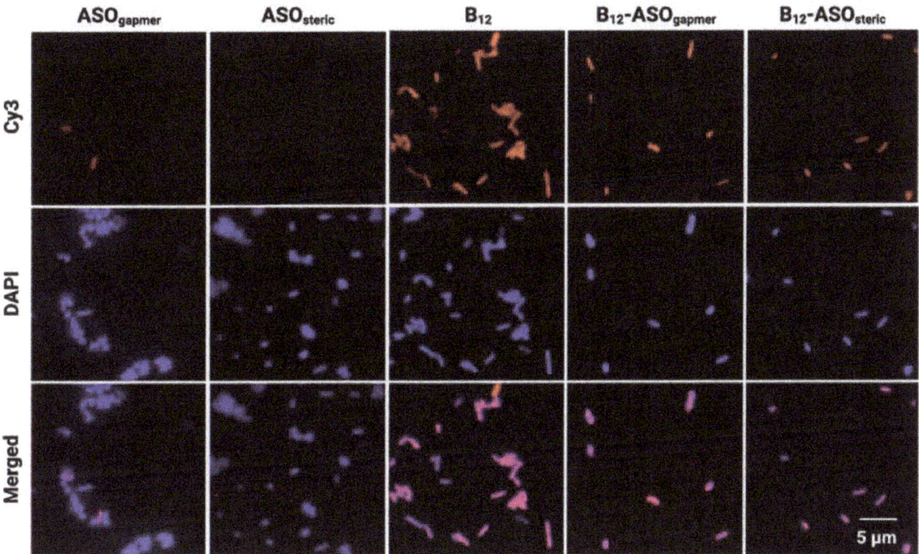

Figure 4. Interaction of Cy3-labeled ASOs, B_{12}, and B_{12} conjugates (concentration of 30 μM) with *E. coli* K12 after 4 h. Bacteria are counterstained with 4′,6′-diamidino-2-phenylindole (DAPI). Images are representative of three independent experiments (using duplicates in each). Scale bar represents 5 μm.

These results point toward the B_{12}-promoted association of the conjugates with the bacterial cells. However, the experimental distinction between membrane-associated and internalized molecules in bacteria remains a difficult task, given the small size of bacteria, which challenges the resolution limit of most standard equipment, including fluorescence microscopes [36].

Therefore, in an attempt to understand if the conjugates were internalized or membrane adhered, as well as to quantify their relative distribution, the bacterial cells were fractionated. A series of washing steps with a gradient of Triton X-100 concentrations was used to differentiate the membrane fraction from the cytosol [37]. These fractions were quantified using a fluorometer. Figure 5 clearly shows that only a small fraction of B_{12} and B_{12} conjugates completely penetrate the bacterial envelope into the cytosol (only 12%, 9%, and 6%, respectively, for the unconjugated B_{12}, B_{12}-ASO_{gapmer}, and B_{12}-ASO_{steric}), while more than 80% remain adhered to the membrane in all cases. The presence on the periplasm is not relevant (only ~3% of the conjugates were retained in this matrix, which was not statistically different from the cytosol. $p > 0.05$, Figure S3 [38]), which indicates that the BtuB at the outer membrane (OM) is likely the limiting factor for conjugate internalization into the cytosol. On the contrary, 4′,6′-diamidino-2-phenylindole (DAPI), a small and cell-permeant DNA intercalating dye, was majorly localized at the cytosol (Figure 5), as expected. Nonetheless, a small fraction was also present in the membrane (Figure 5), which can occur especially in non-fixed cells [39].

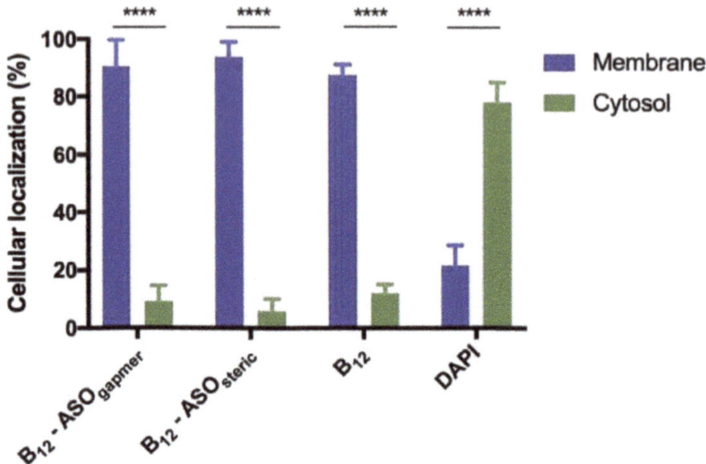

Figure 5. Cellular localization of B_{12} conjugates, B_{12}, and DAPI control in *E. coli* K12. B_{12} conjugates and B_{12} are mainly found on the OM, while the DAPI control is mostly associated with the cytosol. No significant differences were observed between the different internalized conjugates and between the conjugates and the B_{12} control ($p > 0.05$). Significant differences were observed between the membrane and cytosol-associated compounds ($p \leq 0.0001$, ****). The fluorescence of each fraction present in the DAPI control is significantly different from the tested counterparts ($p \leq 0.0001$). Results are presented as mean values and respective standard deviation from three independent assays (using duplicates in each).

From the obtained fractionation results, it can be concluded that the microscopy fluorescence observed in Figure 4 is predominantly derived from conjugates associated with the OM of *E. coli* K12, rather than conjugates internalized in the cytosol, where the ASO would hybridize the *acpP* mRNA target. The inability of B_{12} to serve, in this study, as an efficient trojan-horse for the internalization of ASOs explains the lack of antimicrobial activity of the conjugates observed in Figure 3. It is possible that the uptake of B_{12} is strongly limited by the activity of BtuB, which is present in the OM.

The uptake of B_{12} is regulated by the expression/repression of the BtuB, the locus encoding for the B_{12} receptor [40]. B_{12} acts as a cofactor for methionine synthesis, necessary for growth [41]. In *E. coli*, it has been estimated that the methionine synthesis requires very low levels of B_{12} (20 molecules per cell) [42], while there are hundreds of thousands of mRNA copies of the *acpP* gene [43]. In our work, we have used a much higher concentration of B_{12} than the amount that *E. coli* needs for methionine synthesis. Hence, the difference between the amount of internalized B_{12} conjugates and the high amount of copies of the essential gene we aimed to inhibit may explain the lack of effectiveness of the conjugates.

In addition, it is also important to reflect on the future of this strategy, considering that in vivo, the number of internalized conjugates will probably be even lower since the host cells, as well as other bacteria from the microbiome, will compete for B_{12}. Moreover, most in vivo infections are associated not with single-cell but with clustered cells organized in biofilms [44,45]. Therefore, the bioavailability of B_{12} conjugates may also be limited by interactions with the extracellular matrix. Nonetheless, genes encoding for virulent characteristics, such as the biofilm formation, are usually present in lower amounts of copies. Thus, it would be relevant to study the effect of B_{12}-ASO conjugates targeting these genes in the future. In addition, the in vivo competition for B_{12} will favor bacteria with an improved affinity toward B_{12}, as it has been found for some bacteria in the gut possessing an additional lipoprotein (BtuG) [46]. The use of B_{12} conjugates to target infections caused by such bacteria possessing BtuG could be considered in future biofilm studies.

3. Materials and Methods

3.1. Materials

All basic reagents used were purchased from commercial sources (Sigma-Aldrich, Søhus, Denmark) and used as received. Specific reagents and chemicals included LNA phosphoramidite monomers (Innovassynth Technologies, Maharashtra, India), DNA phosphoramidite monomers (Sigma-Aldrich, St Louis, MO, USA), 2′OMe phosphoramidite monomers, 3′-PT-amino-modifier C6, BCN *N*-hydroxysuccinimide ester (Glen Research, Sterling, VA, USA), DBCO-sulfo-Cy3 (Jena Bioscience, Jena, Germany) and Vitamin B_{12} (Carbosynth, Compton, U.K).

3.2. Synthesis and Design of the ASOs

ASOs were designed to target the gene *acpP*, an essential gene coding for a protein involved in fatty acid biosynthesis. The particular *acpP* target region for the ASOs was selected based on previous studies [22,27,47]. Two different ASOs were synthesized: (i) an LNA/2′OMe chimera, designed for steric blocking (ASO_{steric}), and (ii) an LNA/DNA chimera (ASO_{gapmer}), designed to recruit RNase (sequences are represented in Figure 1a). Since LNA and 2′OMe substitutions increase the duplex stability, the ASO_{steric} was designed to be shorter than the ASO_{gapmer}. ASOs were synthesized under anhydrous conditions using a PerSpective Biosystems Expedite 8909 nucleic acid synthesizer, as described elsewhere [48]. The synthesis was performed on a 1 µmol scale, using a 3′-PT-amino-modifier C6 support, with the following conditions: trichloroacetic acid in CH_2Cl_2 (3:97) as detritylation reagent, 0.25 M 4,5-dicyanoimidazole (DCI) in CH_3CN as an activator, acetic anhydride in THF (9:91, v/v) as cap A solution, *N*-methylimidazole in THF (1:9, v/v) as cap B solution, and a thiolation solution containing 0.0225 M xanthan hydrate in pyridine/CH_3CN (20:90, v/v). The coupling time was 4.6 min for both monomers. To obtain labeled ASOs, Cy3 phosphoramidite was added to the 5′ end in anhydrous CH_3CN (0.1 M) and activated by tetrazole with a 15 min coupling time. The stepwise coupling yields were determined by the UV absorbance (at 500 nm) of dimethoxytrityl cations (DMT^+) that were released after each coupling. The resulting ASOs were purified by reverse-phase HPLC (RP-HPLC), using a Waters System 600 HPLC equipment, equipped with a Waters XBridge BEH C18-column (5 µm, 100 nm × 19 mm). Their composition and purity (>85%) were confirmed by MALDI-TOF MS and ion-exchange HPLC analysis, respectively. Finally, the purified ASOs were labeled by reaction with BCN *N*-hydroxysuccinimide ester I in carbonate buffer (ASO:BCN = 1:2.5 equivalent) for 2 h. BCN labeled ASOs were desalted using NAP-10 Sephadex columns and purified by RP-HPLC. Their composition and purity (>95%) were confirmed by MALDI-TOF MS and analytical reverse-RP-HPLC, respectively. Concentrations of purified oligonucleotides were determined by UV absorption measurements at 600 nm.

3.3. Conjugation of ASOs with Vitamin B_{12}

The 5′-azide-B_{12} was synthesized from commercially available vitamin B_{12} as described by Chromiński et al. [20]. Briefly, the 5′-hydroxy group of the B_{12} was transformed into a good leaving group (a mesyl group), and subsequently, azidation reaction provided the desired 5′-azide-B_{12}. The 5′ position was chosen to avoid obstruction of both components of the conjugate [6]. The azide-B_{12} was isolated through precipitation. MS and NMR data were in accordance with the reported data [20].

Each Cy3-labeled ASO-BCN, dissolved in Milli-Q water, was added to a solution of azido-B_{12}, dissolved in DMSO (ASO: azido-B_{12} = 1:2 equivalent). The resulting solution was transferred to a Biotage microwave reaction vial (0.5 mL) and sealed under a nitrogen atmosphere. The reaction was carried out on a Biotage Initiator microwave synthesizer at 60 °C for 3 h, whereupon all solvents were removed in vacuo, and the residue was re-dissolved in Milli-Q water (Figure 1b). Analytical RP-HPLC and MALDI-TOF MS were performed. The resulting solutions were de-salted by precipitation of the products by first adding an aqueous solution of sodium acetate (3 M, 15 µL) followed by the addition of cold ethanol (1 mL, 99% *w*/*w*; −20 °C). The resulting suspensions were stored at −20 °C for 1 h, and after centrifugation (16,000× *g*, 5 min, 4 °C), the supernatants were removed, and the pellet further washed with cold ethanol (2 × 1 mL; −20 °C), dried for 2 h and then dissolved in Milli-Q water (1 mL). Mass spectra of B_{12}-ASO conjugates were recorded using MALDI-TOF MS, and the purity was confirmed by analytical RP-HPLC. Concentrations of purified conjugates were determined by ultraviolet absorbance at 260 nm.

The same procedure was conducted to obtain fluorescently labeled B_{12} (without ASO) where DBCO-sulfo-Cy3, instead of the Cy3-labeled ASOs, was conjugated to B_{12} through click-chemistry.

3.4. Bacterial Strain and Growth Conditions

E. coli K12 MG1655 was used in this study. To prepare the inoculum, the strain was grown overnight in tryptic soy broth (TSB) at 37 °C with shaking (180 rpm). To monitor both the inhibition of the *acpP* gene expression and the location of the conjugates, *E. coli* K12 was grown in Davis minimal medium at 37 °C with shaking (180 rpm) [49]. This medium lacks B_{12} in its composition, which was crucial to ensure that the internalized B_{12} comes from the control/conjugates incubated with bacteria [6,21].

3.5. Bacterial Susceptibility Tests

The inhibition of the expression of the essential *acpP* gene by the conjugates B_{12}-ASO_{gapmer} and B_{12}-ASO_{steric} was evaluated by monitoring the growth of *E. coli* K12, using a standard microdilution method. An overnight culture of *E. coli* K12 was diluted to an OD_{600} of 0.1 in fresh Davis minimum medium. These cell suspensions were added to wells of sterile 96-well plates and incubated with different concentrations of the tested compounds at 37 °C. The final concentration of the B_{12}-ASOs and respective controls (B_{12}, ASO_{gapmer}, and ASO_{steric}) was 30 µM. As a control for the ASOs activity, the most well-studied vector-ASO conjugate was tested. In brief, a conjugate composed of the cell-penetrating peptide $(KFF)_3K$ and peptide nucleic acid (PNA) (Eurogentec, Seraing, Belgium), designed to hybridize with the same *acpP* sequence was tested in the same conditions as the B_{12} conjugates. The absorbance at 600 nm was determined on a BMGLabtech SPECTROstar Nano microplate reader for 24 h. *E. coli* in medium without any added compound was used as control (CB). Experiments were performed in three independent biological replicates.

3.6. Evaluation of the Internalization of B_{12}-ASOs by Epifluorescence Microscopy

To evaluate the extent of internalized B_{12}-ASO conjugates, compared with the ASOs alone, we used epifluorescence microscopy. An overnight culture of *E. coli* K12 was diluted to an OD_{600} of 0.1 in fresh Davis minimum medium. The B_{12}-ASO_{gapmer}, B_{12}-ASO_{steric}, B_{12}, ASO_{gapmer}, and ASO_{steric} (all Cy3-labeled) were diluted in sterile distilled H_2O and added to the bacterial suspension to a final concentration of 15 and 30 µM per test tube. After 4 h incubation, tubes were centrifuged (3000× *g*, 10 min), and the pellets were resuspended in sterile distilled H_2O. To label the bacterial cytosol, 4′,6′-diamidino-2-phenylindole (DAPI) staining was used. Staining was performed by placing a drop of DAPI (0.5 µg/mL) on top of the dried sample for 5 min. The samples were visualized on a Nikon Eclipse T*i* SR epifluorescence microscope using a Nikon Plan-Apo 100X objective. Ten pictures of each sample were taken randomly, covering all the areas of the sample, using a QImaging Retiga R1 monochromatic camera, and processed with the NIS-Elements Advanced Research. The exposure time and the excitation intensity were maintained throughout the experiments. A G-2A longpass filter (excitation: 535 nm; emission: 580 nm) and a DAPI bandpass filter (excitation: 375 nm; emission: 460 nm) were used. The images obtained using both filters were merged using the Fiji software. Three repeated samples were analyzed for each condition, and three independent experiments were performed.

3.7. Evaluation of the Internalization of B_{12}-ASOs by Bacterial Fractionation

To determine the location of the conjugates in the *E. coli* K12 cells visualized in the previous section and investigate if the observed fluorescence could derive from association to the bacterial envelope, as opposed to intracellular hybridization, the cells were fractionated, and the fluorescence of the outer-membrane fraction and periplasm vs. the fluorescence of the cytosol fraction was measured.

An overnight inoculum of *E. coli* K12 was diluted to an OD_{600} of 0.1 and grown in Davis minimal medium in the presence of 30 μM of B_{12}-ASO_{gapmer}, B_{12}-$ASO_{steric,}$ and B_{12}, (all Cy3-labeled) for 4 h. Thereafter, a fractionation protocol (Figure 6) adapted from Banbula et al. [50] was followed. In brief, bacteria were centrifuged (3000× *g*, 20 min), resuspended in 10 mM Tris-150 mM NaCl (pH 7.4), and washed with 50 mM Tris (pH 7.6). To obtain the fraction associated with the outer-membrane (membrane fraction), bacteria were centrifuged (3000× *g*, 20 min) and resuspended in 50 mM Tris buffer solution containing 0.05% Triton X-100 (pH 7.6), for 1 h at room temperature (RT). After new centrifugation (same conditions), the Cy3 fluorescence intensity of the supernatant (membrane fraction) was measured with a fluorometer (BMGLabtech Fluorostar Omega), using 550 nm excitation and 570 nm emission filters. To obtain the fraction associated with the cytosol, the pellet was resuspended in a more astringent buffer containing 50 mM Tris 1% Triton X-100 (pH 7.6), for 1 h at RT. The supernatant resultant from the last centrifugation was removed, and the Cy3 fluorescence of the cytosol (cytosol fraction) was measured. As a control, the same fractionation protocol was applied to bacteria stained only with DAPI, and the fluorescence of the membrane and cytosol fractions was measured using 375 nm excitation and 460 nm emission filters.

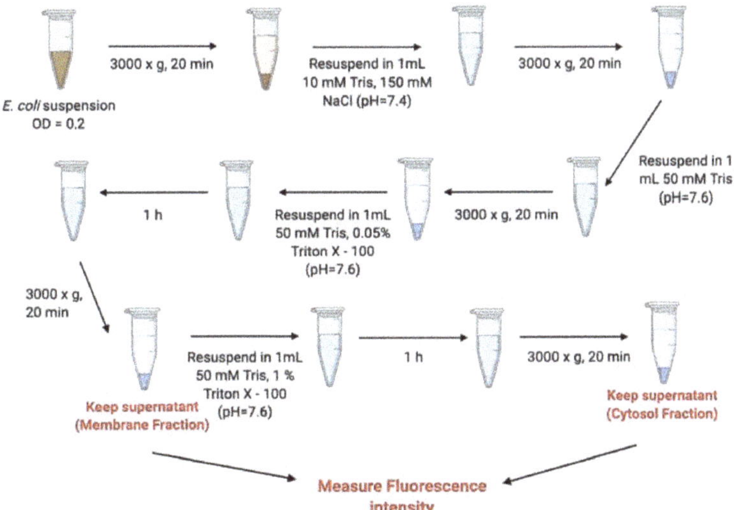

Figure 6. Fractionation protocol adapted from Bandula et al. [50]. A series of washing steps with a Triton X-100 gradient allows the isolation of the membrane and the cytosol fractions.

3.8. Statistical Analysis

For the evaluation of the statistical significance, the two-way analysis of variance test (ANOVA) followed by Sydak's multiple comparisons was used. A *p*-value of $p \leq 0.05$ was considered statistically significant.

4. Conclusions

In summary, this innovative investigation discloses the challenges that need to be overcome before B_{12} -mediated ASO internalization is considered realistic toward tackling the challenge of antimicrobial resistance. This strategy is based on the uptake of the micronutrient B_{12}, which seems to be insufficient to act as an efficient trojan-horse for ASOs designed to inhibit the expression of an essential bacterial gene. In the future, it would be relevant to assess the concentration of internalized ASOs needed to efficiently knock down bacterial genes and inhibit bacterial growth. In addition, improving the bioavailability of vitamin B_{12} by modifying the conjugates and choosing better adapted bacterial targets would be important for successful translation from in vitro to in vivo application.

Supplementary Materials: The following are available online at https://www.mdpi.com/article/10.3390/antibiotics10040379/s1, Figure S1: Analytic RP-HPLC trace and MALDI-MS on B_{12}-ASOgapmer and B_{12}-ASO_{steric}. The top panel shows the retention time (in minutes) of the B_{12} conjugates, and the bottom panel shows the mass (m/z) of each B_{12} conjugate.; Figure S2: Interaction of Cy3 labeled ASOs, B_{12}, and B_{12} conjugates with *E. coli* K12, after 4 h incubation at a concentration of 15 µM. Bacteria are counterstained with DAPI. Images are representative of three independent experiments (using duplicates in each). Scale bar represents 5 µm; Figure S3: Most of the conjugated (B_{12}-ASO_{steric}) and unconjugated B_{12} (B_{12}) are prevented from internalization into *E. coli* cytosol since they remain at the outer-membrane; the percentages in the periplasm are only residual. The isolation of the periplasm was performed after the isolation of the OM fraction and was based on the fractionation protocol by Malherbe et al. 2019. *E. coli* cells were washed in spheroplast buffer (0.1 M Tris-NaCl, 500 mM sucrose, 0.5 mM EDTA, pH 8.0) followed by resuspension in distilled water and incubation for 15 s on ice. The osmotic shock occurred after the addition of $MgSO_4$ (final concentration 20 mM). DAPI was used as a control, majorly localizing at the cytosol, as expected. Statistical differences are indicated when appropriate in * ($p \leq 0.0001$, ****); Table S1: Characterization of the synthesized conjugates, including their HPLC retention times (t_R) and molecular masses, as well as the yield of the respective conjugation reactions.

Author Contributions: Conceptualization, S.P., R.S.S., N.F.A., and J.W.; methodology, S.P., R.Y., P.T.J., J.W., N.F.A., and R.S.S.; experimental, S.P., M.G., and R.Y.; statistical analysis, S.P.; writing—original draft preparation, S.P. and R.S.S.; writing—review and editing, J.W.; N.F.A., P.T.J., M.G., and R.Y.; supervision, R.S.S., N.F.A., and J.W.; funding acquisition, N.F.A. and J.W. All authors have read and agreed to the published version of the manuscript.

Funding: The research was funded by Fundação para a Ciência e Tecnologia, PhD grant SFRH/BD/118018/2016; the Project UID/EQU/00511/2019-Laboratory for Process Engineering, Environment, Biotechnology and Energy–LEPABE funded by national funds through FCT/MCTES (PIDDAC); Project "LEPABE-2-ECO-INNOVATION"–NORTE-01-0145-FEDER-000005, funded by Norte Portugal Regional Operational Programme (NORTE 2020), under PORTUGAL 2020 Partnership Agreement, through the European Regional Development Fund (ERDF); and the European Union's Horizon 2020 research and innovation program under grant agreement No 810685; Biomolecular Nanoscale Engineering Center (BioNEC), a VILLUM center of excellence, funded by VILLUM FONDEN, grant number VKR18333.

Institutional Review Board Statement: Not applicable.

Informed Consent Statement: Not applicable.

Data Availability Statement: The data presented in this study are available in the article and in the supplementary material.

Acknowledgments: Joan Hansen and Tina Grubbe from BioNEC are thanked for technical assistance.

Conflicts of Interest: The authors declare no conflict of interest.

References

1. Baker, S. A return to the pre-antimicrobial era? *Science* **2015**, *347*, 1064–1066. [CrossRef]
2. WHO. *The Evolving Threat of Antimicrobial Resistance: Options for Action*; World Health Organization: Geneva, Switzerland, 2012.
3. Pifer, R.; Greenberg, D.E. Antisense antibacterial compounds. *Transl. Res.* **2020**, *223*, 89–106. [CrossRef]
4. Bai, H.; You, Y.; Yan, H.; Meng, J.; Xue, X.; Hou, Z.; Zhou, Y.; Ma, X.; Sang, G.; Luo, X. Antisense inhibition of gene expression and growth in gram-negative bacteria by cell-penetrating peptide conjugates of peptide nucleic acids targeted to *rpoD* gene. *Biomaterials* **2012**, *33*, 659–667. [CrossRef]

5. Wengel, J. Synthesis of 3 ′-C-and 4 ′-C-branched oligodeoxynucleotides and the development of locked nucleic acid (LNA). *Acc. Chem. Res.* **1999**, *32*, 301–310. [CrossRef]
6. Giedyk, M.; Jackowska, A.; Rόwnicki, M.; Kolanowska, M.; Trylska, J.; Gryko, D. Vitamin B_{12} transports modified RNA into *E. coli* and *S. Typhimurium* cells. *Chem. Commun.* **2019**, *55*, 763–766. [CrossRef]
7. Geary, R.S.; Baker, B.F.; Crooke, S.T. Clinical and preclinical pharmacokinetics and pharmacodynamics of mipomersen (Kynamro®): A second-generation antisense oligonucleotide inhibitor of apolipoprotein B. *Clin. Pharmacokinet.* **2015**, *54*, 133–146. [CrossRef] [PubMed]
8. Kim, J.; Hu, C.; Moufawad El Achkar, C.; Black, L.E.; Douville, J.; Larson, A.; Pendergast, M.K.; Goldkind, S.F.; Lee, E.A.; Kuniholm, A. Patient-customized oligonucleotide therapy for a rare genetic disease. *N. Engl. J. Med.* **2019**, *381*, 1644–1652. [CrossRef] [PubMed]
9. Crooke, S.T. Molecular mechanisms of antisense oligonucleotides. *Nucleic Acid Ther.* **2017**, *27*, 70–77. [CrossRef] [PubMed]
10. Kole, R.; Krainer, A.R.; Altman, S. RNA therapeutics: Beyond RNA interference and antisense oligonucleotides. *Nat. Rev. Drug Discov.* **2012**, *11*, 125–140. [CrossRef] [PubMed]
11. Hegarty, J.P.; Krzeminski, J.; Sharma, A.K.; Guzman-Villanueva, D.; Weissig, V.; Stewart, D.B. Bolaamphiphile-based nanocomplex delivery of phosphorothioate gapmer antisense oligonucleotides as a treatment for *Clostridium difficile*. *Int. J. Nanomed.* **2016**, *11*, 3607. [CrossRef]
12. Turner, J.J.; Ivanova, G.D.; Verbeure, B.; Williams, D.; Arzumanov, A.A.; Abes, S.; Lebleu, B.; Gait, M.J. Cell-penetrating peptide conjugates of peptide nucleic acids (PNA) as inhibitors of HIV-1 Tat-dependent trans-activation in cells. *Nucleic Acids Res.* **2005**, *33*, 6837–6849. [CrossRef]
13. Järver, P.; Coursindel, T.; Andaloussi, S.E.; Godfrey, C.; Wood, M.J.; Gait, M.J. Peptide-mediated cell and in vivo delivery of antisense oligonucleotides and siRNA. *Mol. Ther. Nucleic Acids* **2012**, 1. [CrossRef]
14. Bennett, C.F.; Swayze, E.E. RNA targeting therapeutics: Molecular mechanisms of antisense oligonucleotides as a therapeutic platform. *Annu. Rev. Pharmacol. Toxicol.* **2010**, *50*, 259–293. [CrossRef]
15. Gruber, K.; Puffer, B.; Kräutler, B. Vitamin B_{12}-derivatives—enzyme cofactors and ligands of proteins and nucleic acids. *Chem. Soc. Rev.* **2011**, *40*, 4346–4363. [CrossRef] [PubMed]
16. Bassford, P.; Bradbeer, C.; Kadner, R.J.; Schnaitman, C.A. Transport of vitamin B_{12} in *tonB* mutants of *Escherichia coli*. *J. Bacteriol.* **1976**, *128*, 242–247. [CrossRef] [PubMed]
17. Cadieux, N.; Phan, P.G.; Cafiso, D.S.; Kadner, R.J. Differential substrate-induced signaling through the TonB-dependent transporter BtuB. *Proc. Natl. Acad. Sci. USA* **2003**, *100*, 10688–10693. [CrossRef] [PubMed]
18. Giannella, R.; Broitman, S.; Zamcheck, N. Vitamin B_{12} uptake by intestinal microorganisms: Mechanism and relevance to syndromes of intestinal bacterial overgrowth. *J. Clin. Investig.* **1971**, *50*, 1100–1107. [CrossRef] [PubMed]
19. Wierzba, A.; Wojciechowska, M.; Trylska, J.; Gryko, D. Vitamin B_{12} suitably tailored for disulfide-based conjugation. *Bioconjugate Chem.* **2016**, *27*, 189–197. [CrossRef] [PubMed]
20. Chromiński, M.; Gryko, D. "Clickable" vitamin B_{12} derivative. *Chem. A Eur. J.* **2013**, *19*, 5141–5148. [CrossRef] [PubMed]
21. Rόwnicki, M.; Wojciechowska, M.; Wierzba, A.J.; Czarnecki, J.; Bartosik, D.; Gryko, D.; Trylska, J. Vitamin B_{12} as a carrier of peptide nucleic acid (PNA) into bacterial cells. *Sci. Rep.* **2017**, *7*, 1–11. [CrossRef] [PubMed]
22. Rownicki, M.; Dąbrowska, Z.; Wojciechowska, M.; Wierzba, A.J.; Maximova, K.; Gryko, D.; Trylska, J.J.A.O. Inhibition of *Escherichia coli* growth by Vitamin B_{12}–Peptide Nucleic Acid Conjugates. *ACS Omega* **2019**, *4*, 819–824. [CrossRef]
23. Soares, N.C.; Spat, P.; Krug, K.; Macek, B. Global dynamics of the *Escherichia coli* proteome and phosphoproteome during growth in minimal medium. *J. Proteome Res.* **2013**, *12*, 2611–2621. [CrossRef] [PubMed]
24. Banerjee, D.; Shivapriya, P.; Gautam, P.K.; Misra, K.; Sahoo, A.K.; Samanta, S.K. A review on basic biology of bacterial biofilm infections and their treatments by nanotechnology-based approaches. *Proc. Natl. Acad. Sci.* **2020**, *90*, 243–259. [CrossRef]
25. Howard, J.J.; Sturge, C.R.; Moustafa, D.A.; Daly, S.M.; Marshall-Batty, K.R.; Felder, C.F.; Zamora, D.; Yabe-Gill, M.; Labandeira-Rey, M.; Bailey, S.M. Inhibition of *Pseudomonas aeruginosa* by peptide-conjugated phosphorodiamidate morpholino oligomers. *Antimicrob. Agents Chemother.* **2017**, 61. [CrossRef] [PubMed]
26. Narenji, H.; Teymournejad, O.; Rezaee, M.A.; Taghizadeh, S.; Mehramuz, B.; Aghazadeh, M.; Asgharzadeh, M.; Madhi, M.; Gholizadeh, P.; Ganbarov, K. Antisense peptide nucleic acids against *ftsZ* and *efaA* genes inhibit growth and biofilm formation of *Enterococcus faecalis*. *Microb. Pathog.* **2020**, *139*, 103907. [CrossRef]
27. Good, L.; Awasthi, S.K.; Dryselius, R.; Larsson, O.; Nielsen, P.E. Bactericidal antisense effects of peptide–PNA conjugates. *Nat. Biotechnol.* **2001**, *19*, 360. [CrossRef]
28. Lopez, C.; Arivett, B.A.; Actis, L.A.; Tolmasky, M.E. Inhibition of AAC (6′)-Ib-mediated resistance to amikacin in *Acinetobacter baumannii* by an antisense peptide-conjugated 2′, 4′-bridged nucleic acid-NC-DNA hybrid oligomer. *Antimicrob. Agents Chemother.* **2015**, *59*, 5798–5803. [CrossRef]
29. Azevedo, A.S.; Sousa, I.M.; Fernandes, R.M.; Azevedo, N.F.; Almeida, C. Optimizing locked nucleic acid/2′-O-methyl-RNA fluorescence *in situ* hybridization (LNA/2′OMe-FISH) procedure for bacterial detection. *PLoS ONE* **2019**, *14*, e0217689. [CrossRef]
30. Meng, J.; Wang, H.; Hou, Z.; Chen, T.; Fu, J.; Ma, X.; He, G.; Xue, X.; Jia, M.; Luo, X. Novel anion liposome-encapsulated antisense oligonucleotide restores susceptibility of methicillin-resistant Staphylococcus aureus and rescues mice from lethal sepsis by targeting *mecA*. *Antimicrob. Agents Chemother.* **2009**, *53*, 2871–2878. [CrossRef]

31. Xue, X.-Y.; Mao, X.-G.; Zhou, Y.; Chen, Z.; Hu, Y.; Hou, Z.; Li, M.-K.; Meng, J.-R.; Luo, X.-X. Advances in the delivery of antisense oligonucleotides for combating bacterial infectious diseases. *Nanomed. Nanotechnol. Biol. Med.* **2018**, *14*, 745–758. [CrossRef]
32. Meng, J.; Da, F.; Ma, X.; Wang, N.; Wang, Y.; Zhang, H.; Li, M.; Zhou, Y.; Xue, X.; Hou, Z. Antisense growth inhibition of methicillin-resistant *Staphylococcus aureus* by locked nucleic acid conjugated with cell-penetrating peptide as a novel *FtsZ* inhibitor. *Antimicrob. Agents Chemother.* **2015**, *59*, 914–922. [CrossRef] [PubMed]
33. Lou, C.; Martos-Maldonado, M.C.; Madsen, C.S.; Thomsen, R.P.; Midtgaard, S.R.; Christensen, N.J.; Kjems, J.; Thulstrup, P.W.; Wengel, J.; Jensen, K.J. Peptide–oligonucleotide conjugates as nanoscale building blocks for assembly of an artificial three-helix protein mimic. *Nat. Commun.* **2016**, *7*, 1–9. [CrossRef]
34. Dryselius, R.; Aswasti, S.K.; Rajarao, G.K.; Nielsen, P.E.; Good, L. The translation start codon region is sensitive to antisense PNA inhibition in *Escherichia coli*. *Oligonucleotides* **2003**, *13*, 427–433. [CrossRef] [PubMed]
35. Goltermann, L.; Yavari, N.; Zhang, M.; Ghosal, A.; Nielsen, P.E. PNA length restriction of antibacterial activity of peptide-PNA conjugates in *Escherichia coli* through effects of the inner membrane. *Front. Microbiol.* **2019**, *10*, 1032. [CrossRef] [PubMed]
36. Gottstein, C.; Wu, G.; Wong, B.J.; Zasadzinski, J.A. Precise quantification of nanoparticle internalization. *ACS Nano* **2013**, *7*, 4933–4945. [CrossRef]
37. Clais, S.; Boulet, G.; Kerstens, M.; Horemans, T.; Teughels, W.; Quirynen, M.; Lanckacker, E.; De Meester, I.; Lambeir, A.-M.; Delputte, P. Importance of biofilm formation and dipeptidyl peptidase IV for the pathogenicity of clinical *Porphyromonas gingivalis* isolates. *Pathog. Dis.* **2014**, *70*, 408–413. [CrossRef]
38. Malherbe, G.; Humphreys, D.P.; Davé, E. A robust fractionation method for protein subcellular localization studies in *Escherichia coli*. *Biotechniques* **2019**, *66*, 171–178. [CrossRef]
39. Zweifel, U.L.; Hagstrom, A. Total counts of marine bacteria include a large fraction of non-nucleoid-containing bacteria (ghosts). *Appl. Environ. Microbiol.* **1995**, *61*, 2180–2185. [CrossRef]
40. Kadner, R.J. Repression of synthesis of the vitamin B_{12} receptor *in Escherichia coli*. *J. Bacteriol.* **1978**, *136*, 1050–1057. [CrossRef]
41. Davis, B.D.; Mingioli, E.S. Mutants of *Escherichia coli* requiring methionine or vitamin B_{12}. *J. Bacteriol.* **1950**, *60*, 17. [CrossRef]
42. Di Girolamo, P.M.; Kadner, R.J.; Bradbeer, C. Isolation of vitamin B_{12} transport mutants of *Escherichia coli*. *J. Bacteriol.* **1971**, *106*, 751–757. [CrossRef]
43. Li, G.-W.; Burkhardt, D.; Gross, C.; Weissman, J.S. Quantifying absolute protein synthesis rates reveals principles underlying allocation of cellular resources. *Cell* **2014**, *157*, 624–635. [CrossRef] [PubMed]
44. Schembri, M.A.; Kjærgaard, K.; Klemm, P. Global gene expression in *Escherichia coli* biofilms. *Mol. Microbiol.* **2003**, *48*, 253–267. [CrossRef]
45. Soto, S.; Smithson, A.; Martinez, J.; Horcajada, J.; Mensa, J.; Vila, J. Biofilm formation in uropathogenic *Escherichia coli* strains: Relationship with prostatitis, urovirulence factors and antimicrobial resistance. *J. Urol.* **2007**, *177*, 365–368. [CrossRef] [PubMed]
46. Wexler, A.G.; Schofield, W.B.; Degnan, P.H.; Folta-Stogniew, E.; Barry, N.A.; Goodman, A.L. Human gut Bacteroides capture vitamin B_{12} via cell surface-exposed lipoproteins. *Elife* **2018**, *7*, e37138. [CrossRef] [PubMed]
47. Good, L.; Sandberg, R.; Larsson, O.; Nielsen, P.E.; Wahlestedt, C. Antisense PNA effects in *Escherichia coli* are limited by the outer-membrane LPS layer. *Microbiology* **2000**, *146*, 2665–2670. [CrossRef] [PubMed]
48. Fontenete, S.; Guimarães, N.; Leite, M.; Figueiredo, C.; Wengel, J.; Azevedo, N.F. Hybridization-based detection of *Helicobacter pylori* at human body temperature using advanced locked nucleic acid (LNA) probes. *PLoS ONE* **2013**, *8*, e81230. [CrossRef] [PubMed]
49. Davis, B.D. Isolation of biochemically deficient mutants of bacteria by penicillin. *J. Am. Chem. Soc.* **1948**, *70*, 4267. [CrossRef] [PubMed]
50. Banbula, A.; Bugno, M.; Goldstein, J.; Yen, J.; Nelson, D.; Travis, J.; Potempa, J. Emerging Family of Proline-Specific Peptidases of *Porphyromonas gingivalis*: Purification and Characterization of Serine Dipeptidyl Peptidase, a Structural and Functional Homologue of Mammalian Prolyl Dipeptidyl Peptidase IV. *Infect. Immun.* **2000**, *68*, 1176–1182. [CrossRef]

 antibiotics

Article

Efficacy of Novel Bacteriophages against *Escherichia coli* Biofilms on Stainless Steel

Jean Pierre González-Gómez [1,2], Berenice González-Torres [1,2], Pedro Javier Guerrero-Medina [1], Osvaldo López-Cuevas [2], Cristóbal Chaidez [2], María Guadalupe Avila-Novoa [1,*] and Melesio Gutiérrez-Lomelí [1,*]

1 Centro de Investigación en Biotecnología Microbiana y Alimentaria, Departamento de Ciencias Básicas, División de Desarrollo Biotecnológico, Centro Universitario de la Ciénega, Universidad de Guadalajara, Av. Universidad 1115, Ocotlán 47820, Mexico; jgonzalez.219@estudiantes.ciad.mx (J.P.G.-G.); berenice.gonzalez.220@estudiantes.ciad.mx (B.G.-T.); pjgm@cuci.udg.mx (P.J.G.-M.)

2 Laboratorio Nacional para la Investigación en Inocuidad Alimentaria (LANIIA), Centro de Investigación en Alimentación y Desarrollo, A.C. (CIAD), Carretera a Eldorado Km 5.5, Culiacán 80110, Mexico; osvaldo.lopez@ciad.mx (O.L.-C.); chaqui@ciad.mx (C.C.)

* Correspondence: avila.novoa@cuci.udg.mx (M.G.A.-N.); melesio.gutierrez@academicos.udg.mx (M.G.-L.)

Citation: González-Gómez, J.P.; González-Torres, B.; Guerrero-Medina, P.J.; López-Cuevas, O.; Chaidez, C.; Avila-Novoa, M.G.; Gutiérrez-Lomelí, M. Efficacy of Novel Bacteriophages against *Escherichia coli* Biofilms on Stainless Steel. *Antibiotics* **2021**, *10*, 1150. https://doi.org/10.3390/antibiotics10101150

Academic Editors: Luís Melo and Andreia S. Azevedo

Received: 17 August 2021
Accepted: 22 September 2021
Published: 24 September 2021

Abstract: Biofilm formation by *E. coli* is a serious threat to meat processing plants. Chemical disinfectants often fail to eliminate biofilms; thus, bacteriophages are a promising alternative to solve this problem, since they are widely distributed, environmentally friendly, and nontoxic to humans. In this study, the biofilm formation of 10 *E. coli* strains isolated from the meat industry and *E. coli* ATCC BAA-1430 and ATCC 11303 were evaluated. Three strains, isolated from the meat contact surfaces, showed adhesion ability and produced extracellular polymeric substances. Biofilms of these three strains were developed onto stainless steel (SS) surfaces and enumerated at 2, 12, 24, 48, and 120 h, and were visualized by scanning electron microscopy. Subsequently, three bacteriophages showing podovirus morphology were isolated from ground beef and poultry liver samples, which showed lytic activity against the abovementioned biofilm-forming strains. SS surfaces with biofilms of 2, 14, and 48 h maturity were treated with mixed and individual bacteriophages at 8 and 9 $\log_{10}$ PFU/mL for 1 h. The results showed reductions greater than 6 $\log_{10}$ CFU/cm^2 as a result of exposing SS surfaces with biofilms of 24 h maturity to 9 $\log_{10}$ PFU/mL of bacteriophages; however, the *E. coli* and bacteriophage strains, phage concentration, and biofilm development stage had significant effects on biofilm reduction ($p < 0.05$). In conclusion, the isolated bacteriophages showed effectiveness at reducing biofilms of isolated *E. coli*; however, it is necessary to increase the libraries of phages with lytic activity against the strains isolated from production environments.

Keywords: *E. coli* biofilms; food contact surfaces; biocontrol; bacteriophages

1. Introduction

The meat industry is generally at risk, as many foodborne pathogens such as *Escherichia coli* O157:H7, *Salmonella enterica*, and *Listeria monocytogenes* can form biofilms. The microbial complex communities known as biofilms represent a serious food safety concern [1,2]. Surfaces in meat processing plants have been recognized as an important niche for biofilm formation, and the catalytic reactions that occur during their establishment can damage them. Chemical disinfection is often ineffective for the removal of biofilms due to their matrix, mainly formed from extracellular polymeric substances (EPS), which works as a diffusion barrier, preventing sanitizers from reaching biofilm-forming bacteria [2,3].

Escherichia coli is an abundant bacterium in the production environment of the meat industry. It is usually harmless, although, currently, more than 250 different serogroups of Shiga toxigenic *E. coli* (STEC) have been described and over 150 of these were associated with intra- and extraintestinal consumer diseases [4–6]. Stainless steel (SS) is the main material used as a surface during the slaughtering and manipulation of meat. However, *E. coli* is capable of forming biofilms on it, so from these biofilms, viable pathogens could become detached and lead to cross-contamination [7,8]. Thus, due to the high *E. coli* incidence in meat processing plants and the poor accessibility to and difficulty of regular cleaning and disinfection procedures, surface biofilms may pose a food safety concern.

Biofilms play a protective role for bacteria against chemical disinfection in meat processing plants. Thus, the use of bacteriophages (phages) to eliminate biofilms is a promising approach, as they show interesting properties in terms of biofilm removal through the production of enzymes that allow them to actively penetrate and disrupt biofilms [9]. These viruses can infect bacteria following lytic or lysogenic cycles. The lytic cycle ends with the lysis of the host and the release of progeny ready to infect the surrounding bacteria [1]. The use of bacteriophages as an additive in beef and poultry products was approved by the FDA in 2006 [10], and the potential of phages as a biocontrol method became popular after this event, giving rise to multiple studies into the elimination or reduction of both the pathogenic and spoilage bacteria that can be found within the food industry in biofilms or planktonic cells [11–14].

The interaction between bacteriophages and biofilms was described even before they were approved as additives. Recent studies have reported that several factors influence biofilm reduction, such as the ability of the strain to form biofilms, the biofilm development stage, the affinity of the bacteriophages for the strains that produce the biofilm, and the concentration of the bacteriophage, as well as whether it is applied individually or as a mixture [1,15–17]. The aim of this study was to isolate, characterize, and challenge natural lytic bacteriophages against biofilm-forming *E. coli* strains isolated from the meat industry and to evaluate the main factors that influence their effectiveness in the removal of biofilms developed on stainless steel surfaces.

2. Results

2.1. Biofilm Formation Ability of E. coli Strains

All the *E. coli* strains, isolated from the meat industry and ATCC, produced black colonies with a dry and crystalline consistency in Congo Red Agar (CRA) and were recorded as EPS producers. Whereas in the semiquantitative adherence test, MGA-EC-25 and MGA-EC-27 showed a strong adhesion ability with the highest ODs (0.898 ± 0.113 and 0.968 ± 0.042, respectively), MGA-EC-21 and *E. coli* ATCC 11303 showed a weak adhesion ability (0.095 ± 0.021 and 0.174 ± 0.018, respectively), while the rest of the strains were recorded as having null adhesion ability (Table 1).

Table 1. Characterization of the biofilm formation ability of the *E. coli* strains used in this study.

Bacterial Strain	Adherence Assay		Phenotype CRA [b]
	OD (λ = 570)	Adhesion Ability [a]	
MGA-EC-01	0.076 ± 0.008	Null	EPS producer
MGA-EC-02	0.073 ± 0.008	Null	EPS producer
MGA-EC-08	0.066 ± 0.004	Null	EPS producer
MGA-EC-21	0.095 ± 0.021	Weak	EPS producer
MGA-EC-23	0.067 ± 0.005	Null	EPS producer
MGA-EC-25	0.898 ± 0.113	Strong	EPS producer
MGA-EC-26	0.093 ± 0.026	Null	EPS producer
MGA-EC-27	0.968 ± 0.042	Strong	EPS producer
MGA-EC-28	0.071 ± 0.007	Null	EPS producer
MGA-EC-30	0.062 ± 0.001	Null	EPS producer
ATCC BAA-1430	0.074 ± 0.003	Null	EPS producer
ATCC 11303	0.174 ± 0.018	Weak	EPS producer

[a] Null adherent ability: OD ≤ 0.093; weak adhesion ability: 0.093 < OD ≤ 0.186; moderate adhesion ability: 0.186 < OD ≤ 0.373; strong adhesion ability: 0.373 < OD. [b] EPS producer: black colonies of dry crystalline consistency in CRA; EPS nonproducer: red colonies in CRA.

2.2. Biofilm Development Curve

MGA-EC-21, MGA-EC-25, and MGA-EC-27 were selected for determination of the biofilm formation curve, the *E. coli* strain ATCC BAA-1430 was included as a surrogate indicator, and *E. coli* ATCC 11303 was included as a positive control for its reported biofilm-forming ability. The biofilms' cell densities are summarized in Figure 1. The results showed that the three strains isolated from the meat industry reached a higher cell density in the early stages of biofilm formation compared with the ATCC; however, at 120 h, the densities reached by all the strains were between 7 and 8 $\log_{10}$ CFU/cm^2. Moreover, in the strains from the meat industry, we observed that the highest bacterial counts were reached at 12 h, followed by the typical detachment phase of biofilms at 24 h; subsequently, the cell densities remained constant until 120 h, related to biofilm establishment. Conversely, the ATCC strains showed a constant increase in biofilm cell density at the early stages, reaching concentrations similar to those isolated from the meat industry at 120 h.

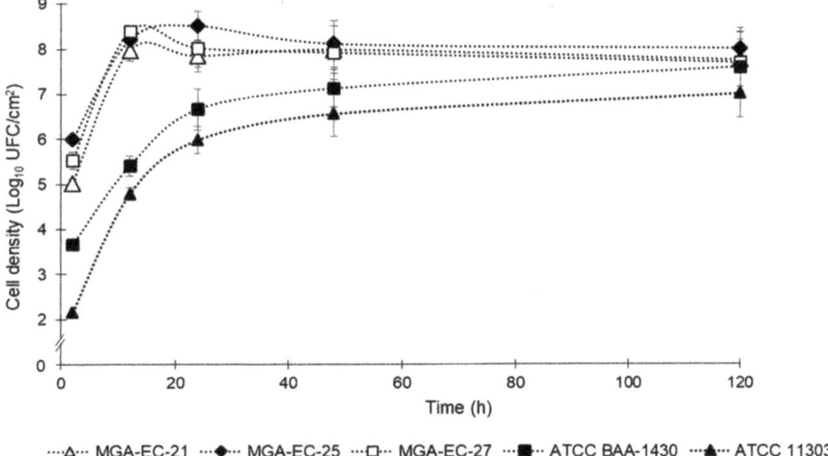

Figure 1. Biofilm development curve for SS coupons of the strains MGA-EC-21, MGA-EC-25, MGA-EC-27, *E. coli* ATCC BAA-1430, and *E. coli* ATCC 11303, at 2, 12, 24, 48, and 120 hours. Values are the means of three tests and vertical bars represent the standard deviations.

2.3. Bacteriophages Isolation and Characterization

Samples of poultry liver and ground beef were collected for the isolation of bacteriophages with lytic activity against biofilm-forming *E. coli* strains from the production environment of the meat industry. Three bacteriophages were isolated, purified, and named according to ICTV recommendations: vB_EcoP_PL-01, vB_EcoP_GB-02, and vB_EcoP_GB-03; their characterization is described in Table 2. In the host range determination, phages PL-01, GB-02, and GB-03 only produced lysis zones against the three strains from the meat industry that showed better biofilm formation ability, while none of the phages lysed any of the ATCC strains. All bacteriophages produced clear, round plaques with diameters of approximately 3 to 4 mm on the lawn of MGA-EC-27. Micrographs obtained from PL-01, GB-02, and GB-03 revealed similar podovirus morphologies with isometric heads 62.20, 51.10, and 49.39 nm in diameter, respectively, and short noncontractile tails (Figure 2).

Table 2. Characterization of bacteriophages with lytic activity against biofilm-forming *E. coli* strains.

Phage Strain	Morphology [a]	Source	Host Range (MGA-EC Strains) [b]										Plaque Diameter [c]
			01	02	08	21	23	25	26	27	28	30	
PL-01	Podovirus	Poultry liver	-	±	±	++	-	++	-	++	-	-	3 mm
GB-02	Podovirus	Ground beef	±	-	-	+	-	+	-	+	-	+	4 mm
GB-03	Podovirus	Ground beef	±	-	-	++	-	++	-	++	-	±	3.5 mm

[a] Classification according to the morphological characteristics of TEM micrographs following the guidelines of the International Committee on Taxonomy of Viruses. [b] Host range results were recorded as follows: clear zone of complete lysis: ++; clear zone of lysis: +; incomplete lysis: ±; no lysis: -. [c] Plaque diameter obtained via the double agar technique using the MGA-EC-27 strain as the host.

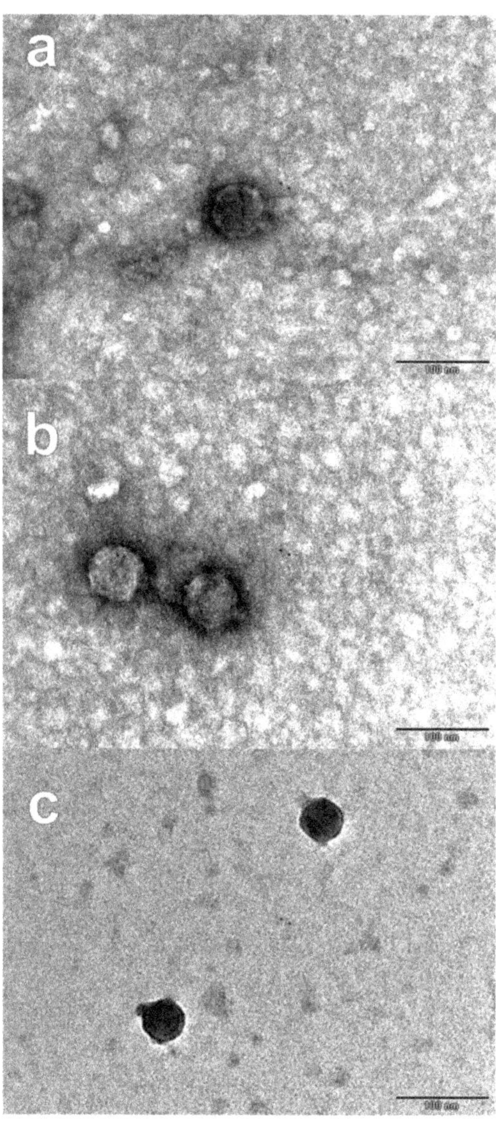

Figure 2. TEM micrographs of bacteriophages PL-01 (**a**), GB-02 (**b**), and GB-03 (**c**) isolated from poultry liver and ground beef.

2.4. Biofilm Reduction Efficacy of Bacteriophages

Biofilms of three *E. coli* strains at different maturity stages were exposed to treatment with three bacteriophages individually and in a mixture at two concentrations. The results showed that depending on the *E. coli* strain, the biofilm maturity, the phages' formulation, or the concentration applied, a statistically significant effect in reducing the cell density of the biofilm could be observed ($p < 0.05$). Afterward, an ANOVA was performed for each bacteriophage concentration because this was the factor with the greatest influence on the removal of biofilms. The treatments with bacteriophages at 10^9 PFU/mL showed the greatest reductions, ranging from 2.39 to 6.79 $\log_{10}$ CFU/cm^2 and reaching greater reductions at 24 and 48 h of biofilm maturity (Figure 3). Interestingly, the individual phages and the mixture had no noticeable differences in their effects. Furthermore, the reductions observed with treatments with phages at 10^8 PFU/mL ranged between 0.95 and 2.86 $\log_{10}$ CFU/cm^2, and the reduction effects at the different development stages were similar. The greatest reduction, 6.70 $\log_{10}$ CFU/mL, was observed when the biofilm formed by MGA-EC-25 at 24 h of maturity was exposed to phage GB-03 at a concentration of 10^9 PFU/mL.

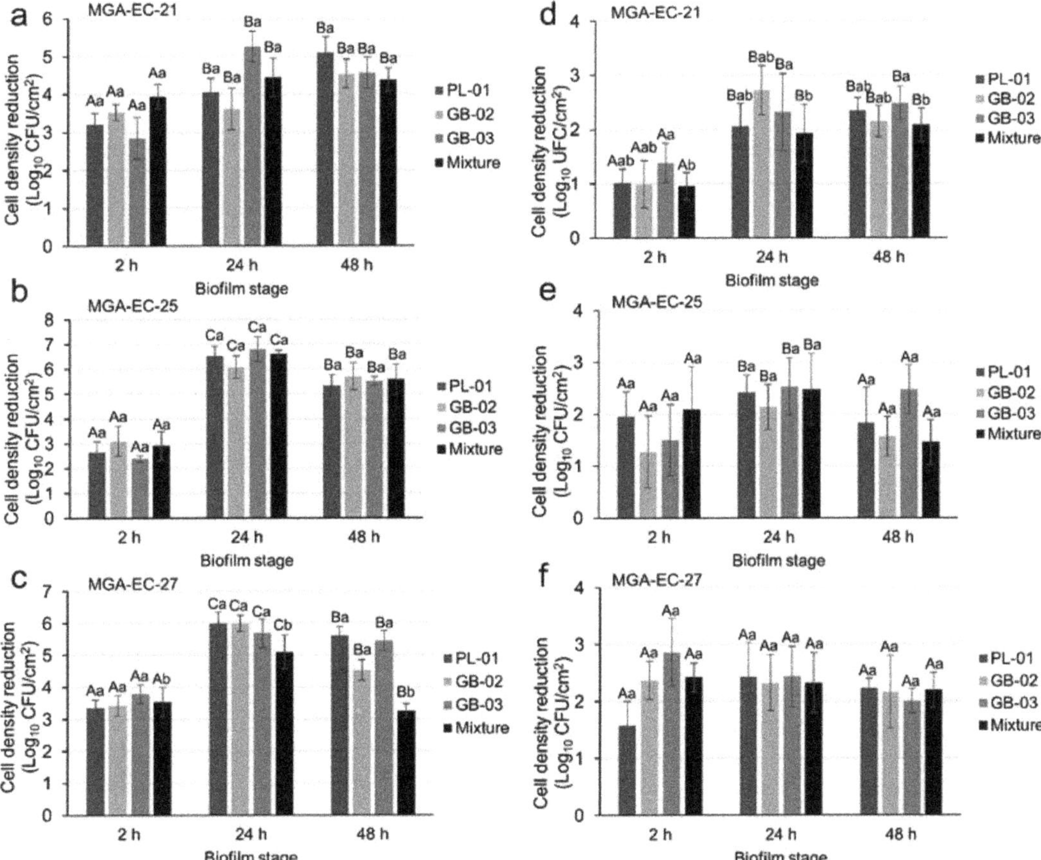

Figure 3. Effect of bacteriophages at concentrations of 10^9 (**a–c**) and 10^8 (**d–f**) PFU/mL for 1 h on MGA-EC-27, MGA-EC-25, and MGA-EC-21 biofilms at 2, 24, and 48 h of maturity. Different uppercase letters indicate a significant difference ($p < 0.05$) in the maturity of the biofilm and lowercase letters indicate a significant difference in the phage or mixture of phages applied. Values are the means of three replicates ± standard deviation.

2.5. Scanning Electron Microscope Analysis

SEM micrographs showed different cell densities and morphologies at different maturity stages in *E. coli* biofilms, as summarized in Figure 4. The micrographs of the untreated surfaces showed that the biofilm density varied in the different stages of development. At 2 h of development, small groups of cells with a well-defined morphology adhering to the SS coupon were observed, while at 24 h, cell agglomerations with higher density were observed and the cell boundaries were not well defined in some regions due to the early production of extracellular polymeric substances. After 48 h of development, the biofilms showed a more compact structure in the central regions and better defined cells towards the edges; furthermore, the cells embedded in EPS formed three-dimensional structures typical of mature biofilms. Micrographs of the bacteriophage-treated surfaces showed similar results at different stages of biofilm development, with altered morphology in most of the cells due to the bacteriophages' lytic activity and some intact cells, possibly due to the generation of phage-resistant cells or the bacteriophage failing to reach its receptor and infect the cell.

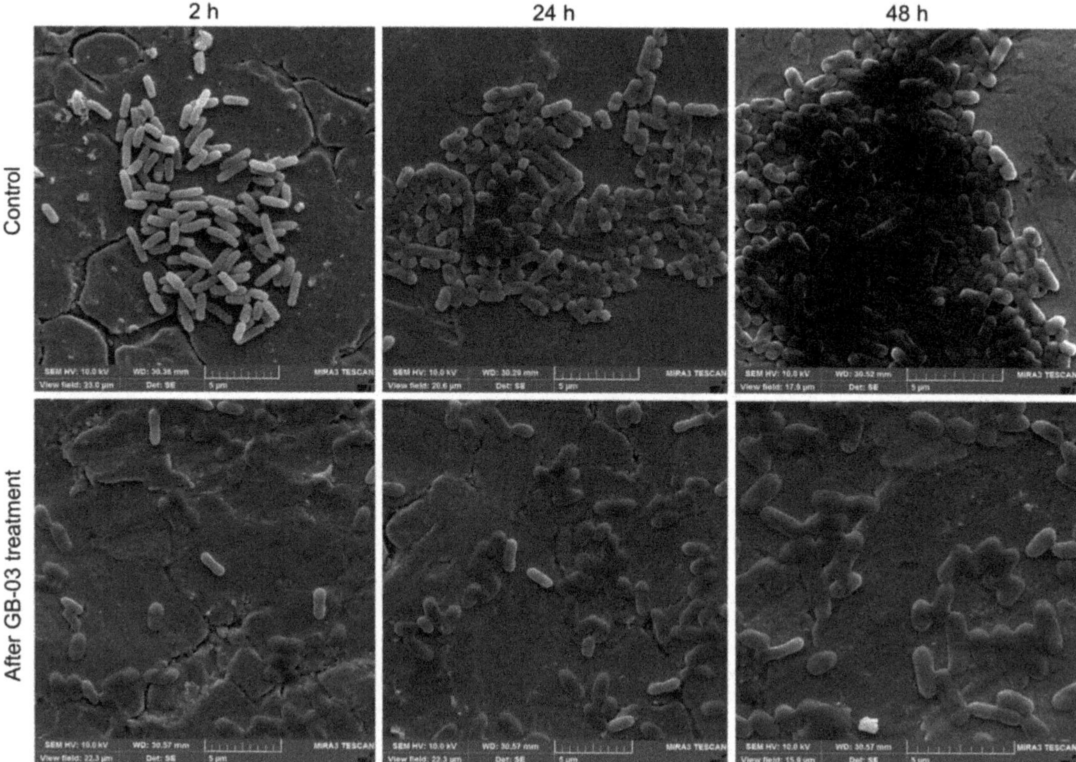

Figure 4. SEM micrographs of MGA-EC-25 biofilms developed on SS coupons at 2, 24, and 48 h with no treatment and after GB-03 exposure at 10^9 PFU/mL for 1 h.

3. Discussion

Biofilms formed on food contact surfaces can reduce the effectiveness of sanitizers, damage the equipment, and contaminate the food product, which may lead to significant public health problems [18]. This study was conducted to evaluate the biofilm-forming ability of *E. coli* strains present in the meat industry environment and to evaluate the efficacy of novel bacteriophages as a potential biocontrol method to remove *E. coli* biofilms. Our results indicate the presence of strong biofilm-forming *E. coli* strains in the meat industry production environment. The Congo Red agar method showed that 10 (100%) *E. coli* strains from the meat industry were positive for EPS production; however, only two (20%) strains showed strong adhesion ability and one (10%) showed weak adhesion ability. The strain *E. coli* ATCC 11303, previously described as biofilm-forming [15], was also positive to EPS production and showed a weak adhesion ability. In similar studies, different proportions of EPS-producing strains and adherence ability have been found. Onmaz et al. [19] reported that 24% of *E. coli* strains isolated were EPS producers, and 36% showed at least weak adhesion ability. The majority of studies were based solely on using the adhesion ability to classify the strains as biofilm-forming [6,20–22]; however, we recommend performing both tests to have a better characterization of the main factors that lead the strains to produce biofilms. These studies reported results ranging from 11% to 100% of strains that showed some adhesion ability, suggesting that this characteristic is highly variable within the same species, which may be due to differences in flagellar motility, which helps the bacteria to counteract the electrostatic and hydrodynamic forces near surfaces, or it may depend on the type of surface because adhesion to abiotic surfaces is mediated by electrostatic and physicochemical interactions among the bacterial membrane, the surfaces, and the medium to which they are exposed [23,24]. Interestingly, the MGA-EC-25 and MGA-EC-27 strains showed 0.898 ± 0.113 and 0.968 ± 0.042 OD, respectively; these values are significantly higher than those for the EMC17 strain (approximately 0.43 OD) [20], the highest value reported in previous studies to our knowledge.

The *E. coli* biofilm stages on SS were observed in the development kinetics and the SEM micrographs. In the first 2 hours of development, relatively low counts were obtained in all evaluated strains, which could be due to the absence of organic matter, which enhances the initial adhesion through preconditioning the surface [25]. At 12 h, the counts increased considerably in the three strains obtained from the meat industry, reaching densities of approximately 8 $\log_{10}$ CFU/cm^2, while both ATCC strains also increased, reaching approximately 5 $\log_{10}$ CFU/cm^2, showing the highest point of the reversible initial adsorption. The cell desorption phase was observed at 24 h of development, when a decrease in the counts of the meat industry strains was obtained, while the ATCC strains continued slower but constant development of the biofilms. The following phases of biofilm formation (irreversible adhesion, microcolony formation, and maturation) were observed in the kinetics as stabilization of the cell density up to 128 h. These results are similar to those obtained by Carson et al. [26], who used the ATCC 11303 strain for an evaluation of biofilm development on polycarbonate, showing cell densities of 6 $\log_{10}$ CFU/well at 24 h of development, close to the 5.98 ± 0.30 $\log_{10}$ CFU/cm^2 obtained in this study, notwithstanding that the surfaces were different. The surrogate strain ATCC BAA-1430 is recommended for the validation of disinfection methods and critical control points in production environments [27]; however, the results show that the strains from the meat industry had a greater ability to form biofilms than this surrogate; therefore, these data should be considered to choose a representative strain to evaluate disinfectants for biofilm removal. In this study, *E. coli* strains isolated from the meat industry reached considerably higher biofilm populations than those reported in similar work; however, most previous research used ATCC strains for the evaluation of biofilms, and information on strains isolated from production environments is scarce. Wang et al. [28] evaluated *E. coli* O145:H25 EC19990166 biofilms on SS surfaces; after 24 and 48 h of development, the biofilms showed populations of 4.7 ± 0.2 and 5.4 ± 0.2 $\log_{10}$ CFU/coupon (~8 cm^2), respectively. Likewise, Kang et al. [29] evaluated the multistrain biofilm of *E. coli* O157:H7 ATCC 8624, 2026, and 2029 after 5 days of development and obtained populations of 5.90 $\log_{10}$ CFU/cm^2. In this study, we obtained populations of approximately 8 $\log_{10}$ CFU/cm^2 in the biofilms of three strains of *E. coli* from 24 to 120 h, with slight fluctuations, that is, 2 to 3 $\log_{10}$ more than in previous studies. Therefore, it is necessary to characterize a greater number of strains from production environments since, as our results show, their biofilm formation capacity is greater than the strains that belong to a collection, and thus, strains from production environments should be used to evaluate disinfection methods. In addition, it should be considered that in real production environments, biofilms are formed by multiple strains, species, and bacterial genera, aside from the different temperatures, humidity, organic matter, and other factors that could affect the population and composition of the biofilm [25,30].

In this study, the evaluation of three novel bacteriophages against *E. coli* biofilms was conducted. Based on the morphological properties of phages PL-01, GB-02, and GB-03, they would have been classified in the order *Caudovirales* within the family *Podoviridae* [31]. However, in recent years, the taxonomy of viruses has changed and now nine families are recognized within the *Caudovirales* order and, in addition to their morphology, their genomic characteristics must be considered for a correct classification [32]. These bacteriophages showed affinity against strains with a greater ability to form biofilms, probably due to the presence of a specific receptor, which could be verified by further studies on phage-resistant bacteria to analyze if this receptor has been lost and how this affects the ability of biofilm formation, as reduced virulence has been shown in other bacteria after losing the phage-specific receptor (capsular polysaccharides, teichoic acids, and pilus) [33]. Both individual phages and phage cocktails were challenged against biofilms of three *E. coli* strains at 2, 24, and 48 h of development, showing results ranging from 0.95 $\log_{10}$ CFU/mL to 6.70 $\log_{10}$ CFU/mL of biofilm reduction after 1 hour of treatment; greater efficacy was observed when applying the treatment at a concentration of 10^9 PFU/mL at 24 and 48 h of biofilm development. Interestingly, the individual phages showed reductions equal or greater than the cocktail in most treatments. Similar results were reported by Montso et al. [21], who reported that the individual phages obtained greater efficacy than the cocktails after 1 day of treatment; nevertheless, the authors observed that after 7 days, there was bacterial regrowth in the treatments with individual phages but not in the cocktails. In addition to being effective at preventing bacterial growth, phage cocktails can be used to overcome the generation of phage resistance and to expand the range of bactericidal action [34,35]. Other studies have reported reductions ranging from 2.9 $\log_{10}$ CFU/coupon [28] and 3.8 $\log_{10}$ CFU/cm^2 [36] to 4.5 $\log_{10}$ CFU/blade [7], which highlights the great efficacy of the phages isolated in this study at reducing *E. coli* biofilms, especially at 24 and 48 h of development.

Bacteriophages as a biofilm biocontrol method have great potential, but it is necessary to fully understand the phage–biofilm interaction to implement all possible improvements. The first important step for phage infection is the adsorption to its receptor, and the biofilm matrix represents a barrier between the phage and its receptor. Multiple studies have shown that some phages possess depolymerase polysaccharides in the spike or tail spicule proteins, and these enzymes allow phages to degrade the polysaccharides that form the extracellular matrix and facilitate their dispersion through the biofilm [1,37,38]. Furthermore, bacteriophages having a podovirus (short tail) morphology can diffuse better through the biofilm compared with siphoviruses and myoviruses. The enzymes present in the phages have been studied for their potential to be used individually for the removal of biofilms, since they have advantages over bacteriophages such as their greater host range, there is no risk of transferring virulence genes, and no resistant bacteria are produced [1]. The lytic activity shown by the bacteriophages PL-01, GB-02, and GB-03 specifically against the *E. coli* strains with the highest biofilm formation ability and high rates of biofilm reduction may be due to binding to a receptor that is involved in the formation of EPS and the presence of enzymes that allow the degradation of the extracellular matrix [1]. To achieve a better understanding of these interactions, it is necessary to perform the sequencing of the phages, *E. coli* strains, and phage-resistant strains to determine the specific receptor of these phages and to investigate their role in the formation of biofilms, as well as to obtain and purify the phages' proteins to characterize their depolymerase activities.

4. Materials and Methods

4.1. Bacterial Strains

The *E. coli* strains used in the present study were isolated from the surfaces of meat processing plants. These strains were kindly provided by the Microbiology Laboratory of Cuciénega UDG (Table 1). Additionally, the surrogate indicator strain *E. coli* ATCC BAA-1430 was used as a reference strain, and *E. coli* ATCC 11303 was used as a positive control for its reported biofilm formation ability. All the strains were subcultured in tryptic soy broth (TSB; Becton Dickinson Bioxon, Le Pont de Claix, France) for 24 h at 37 °C to obtain a final concentration of 10^8 CFU/mL.

4.2. Characterization of the Strains' Biofilm-Forming Ability

4.2.1. Production of Extracellular Polymeric Substances (EPS)

EPS production was evaluated according to the CRA method described by Mariana et al. [39], with some modifications. The test was carried out with two formulations: the first was prepared with a blood agar base, with 0.4 g/L Congo Red and 36 g/L glucose added; in the second formulation, glucose was replaced by 36 g/L sucrose. *E. coli* strains were inoculated in the medium and incubated under aerobic conditions for 48 h at 37 °C. The strains that produced black colonies with a dry crystalline consistency were recorded as EPS producers, while those that grew as red colonies were recorded as nonproducers.

4.2.2. Semiquantitative Adherence Assay

The ability of the strains to adhere to abiotic surfaces was evaluated in 96-well flat-bottomed microtiter polystyrene plates according to the method described by Milanov et al. [40], with some modifications. For each strain, 3 wells of the microtiter plate were filled with 200 µL of a bacterial suspension in TSB with 0.5% glucose (*w/v*) (TSB + G), 3 wells were filled with the TSB + G and used as negative controls, and *E. coli* ATCC 11303 was used as the positive control. The plates were then incubated at 37 °C for 24 h. Briefly, the contents of the wells were removed by inverting the plates, and each well was washed 3 times with phosphate-buffered saline (PBS; 7 mM Na_2HPO_4, 3 mM NaH_2PO_4, and 130 mM NaCl; pH 7.4) to remove the planktonic bacteria. The attached bacteria were fixed with 100 µL of 95% ethanol for 5 min and the plates were emptied and left to dry. Staining was performed with 100 µL of 1% crystal violet for 5 min, then 3 washes with PBS were carried out, and the plates were allowed to dry at room temperature. The plates were stained with 100 µL of 1% (*w/v*) crystal violet solution per well for 5 min. The excess stain was rinsed off with sterile distilled water, and the microtiter plates were air-dried. Optical density (OD) was measured at λ = 570 nm using the Multiskan FC (Thermo Fisher Scientific Inc., Madison, WI, USA). The cutoff value of OD (ODc) was defined as three standard deviations above the mean OD of the negative control. The strains were classified as having null adherent ability (OD $\leq$ ODc), weak adhesion ability (ODc < OD $\leq$ 2 × ODc), moderate adhesion ability (2 × ODc < OD $\leq$ 4 × ODc), or strong adhesion ability (4 × ODc < OD).

4.3. Biofilm Formation on Stainless Steel

4.3.1. Biofilm Quantification

The biofilm-formation ability of *E. coli* was investigated on stainless steel (SS; AISI 316, 8 × 20 × 1 mm) coupons. SS coupons, previously treated and sterilized, were placed individually into the glass test tubes (20 × 150 mm) containing 5 mL of TSB + G [41,42]. For each strain, 5 tubes were inoculated with 50 µL of the bacterial culture (10^8 CFU/mL) and were incubated at 22 ± 2 °C for 2, 12, 24, 48, and 120 h, respectively. After incubation, SS coupons were removed under sterile conditions using sterile forceps and rinsed 2 times by pipetting 1 mL of PBS, and placed independently in tubes containing 9 mL of casein peptone (BD, Bioxon, Becton Dickinson, Le Pont de Claix, France), and the biofilms were removed by sonication (50–60 Hz for 1 min; Sonicor Model SC-100TH, West Babylon, NY, USA). Serial dilutions and conventional plating on tryptic soy agar (TSA; Becton Dickinson, Le Pont de Claix, France) were used to estimate the viable cells in the biofilm. The plates were incubated at 37 °C for 24 h. Three replicates were performed for each strain, and an SS coupon without inoculum was included in all assays as a negative control.

4.3.2. Scanning Electron Microscopy

After each incubation period, the coupon was removed from the tube, rinsed with PBS, and immersed in 2% glutaraldehyde at 4 °C for 2 h to fix the adhering bacteria [43]. Briefly, the SS coupons were vacuum-dried and gold-coated for 30 s [44]. Biofilms were observed by using a TESCAN Mira3 LMU scanning electron microscope (Brno-Kohoutovice, Czech Republic).

4.4. Bacteriophage Isolation

Twenty samples of ground beef and poultry liver were collected from the municipal market of Ocotlán, Jalisco, Mexico. Three milliliters of an overnight culture of each of the 10 previously isolated *E. coli* strains grown in TSB was mixed with 5 g of ground beef or chicken liver in a sterile 50 mL conical tube. The enriched samples were incubated for 24 h at 37 °C and 70 rpm in a shaking bath. The tubes were centrifuged at 10,000× *g* for 10 min at 4 °C (Megafuge 16R, Thermo Fisher Scientific Inc., Waltham, MA, USA), and the supernatant was filtered twice through a sterile nitrocellulose membrane (0.45 and 0.22 μm pore diameter, respectively), using a vacuum pump. The filtered samples (lysates) were used to perform the SPOT test, through the soft overlay technique with 0.4% agarose, against 10 *E. coli* strains [45,46]. The soft agar technique, which involved mixing 1 mL of the overnight cultures with 100 μL of the filtrates that showed lytic activity, was used to observe the production of plaques. Plaques were selected on the basis of size and clarity and transferred to microtubes containing 1 mL of nanopure water. The procedure was repeated at least 3 times per sample to obtain purified phages [7,15,40,47,48].

4.4.1. Phage Host Range

Ten *E. coli* strains were used to test the infection spectrum of the isolated phages. The bacterial strains were cultured in TSB at 37 °C overnight with constant shaking (70 rpm), and 1 mL of each strain was mixed with 3 mL of 0.4% top agarose at 45 °C. Briefly, the suspension was poured into a petri dish with TSA and solidified at room temperature. Next, 10 μL of each phage suspension (10^8–10^9 PFU/mL) was spotted on the soft agar overlay, left to dry, and incubated for 18–24 h at 37 °C [46]. The results were interpreted and recorded as follows: a clear zone of complete lysis: ++; incomplete lysis: +; no lysis: - [49].

4.4.2. Phage Morphology Determined by Transmission Electron Microscopy

The 3 isolated phages were examined by transmission electron microscopy (TEM). A drop of high-titer phage stock (approximately 10^9 PFU/mL) was placed on the surface of a formvar-coated grid (400 mesh copper grid), negatively stained with 2% phosphotungstic acid (pH 7.2) for 5 min, and the excess was removed with filter paper. The grid was carbon-shadowed in a vacuum evaporator (JEOL, JEE400). Electron micrographs were taken at various magnifications in a JEOL JEM-1011 transmission electron microscope [49].

4.5. Biofilm Exposure to Bacteriophages

Biofilms of the *E. coli* strains that showed greater adherence ability (MGA-EC-21, MGA-EC-25, and MGA-EC-27) were promoted on SS coupons as described in Section 4.3.1. After 2, 24, and 48 h of incubation time, the coupons were removed under sterile conditions and washed with 1 mL of PBS to remove planktonic cells. Each SS coupon was deposited in 3 mL of the bacteriophage solution (individual or mixed) at concentrations of 10^8 or 10^9 PFU/mL and exposed for 1 h. The coupon was then extracted from the phage solution and washed with 1 mL of PBS, and the biofilms were removed by sonication in a tube with 9 mL of casein peptone. Serial dilutions and standard plates on TSA were used to estimate viable cells in the biofilm after exposure to phages, and the results obtained were compared with the biofilm development curve of each strain to obtain the cell density reduction [7,14,30]. Three replicates were performed per treatment.

4.6. *Statistical Analysis*

All the experiments were performed in triplicate, and the data were evaluated using analysis of variance (ANOVA), followed by a least significant difference (LDS) test, in Statgraphics Centurion XV software v15.2.06 (Statpoint Technologies, Inc., Warrenton, VA, USA).

5. Conclusions

The present study showed that *E. coli* strains isolated from the meat industry are capable of producing EPS, and some of them showed the ability to adhere to surfaces and produced mature biofilms on stainless steel at 48 h of development. Furthermore, the three isolated bacteriophages showed affinity against the strains with the highest biofilm formation capacity and showed efficient lytic activity against *E. coli* biofilms, mainly at 24 and 48 h of maturity. Interestingly, the application of the bacteriophages in a mixture did not show a greater efficiency compared with the application of individual phages; in some treatments, the effectiveness of the individual application was significantly higher than that of the phage mixture. Nevertheless, the application of mixtures of bacteriophages has multiple advantages, such as the wide range of strains that can be infected or the generation of bacteria that are resistant to one of the bacteriophages but can be infected by another phage in the mixture. Bacteriophages have the potential to be used as a biocontrol method against *E. coli* biofilms in the meat industry, reducing the risk of product contamination and avoiding the deterioration of equipment and surfaces. However, it is necessary to characterize the biofilm formation ability of the strains of interest and to increase the libraries of phages that show specific activity against the biofilm-producing strains, in addition to characterizing the phage enzymes that can also be used for the removal of biofilms. The enzymes present in the phages have been studied for their potential to be used individually for the removal of biofilms, since they have advantages over bacteriophages, such as the greater range of hosts, there is no risk of transferring virulence genes, and no resistant bacteria are produced.

Author Contributions: Conceptualization, methodology, validation, formal analysis, investigation, data curation, writing—original draft preparation, and visualization: J.P.G.-G.; methodology and investigation: B.G.-T.; investigation, validation, and formal analysis: P.J.G.-M., O.L.-C. and C.C.; writing—review and editing, supervision, resources, project administration, funding acquisition, and visualization: M.G.A.-N. and M.G.-L. All authors have read and agreed to the published version of the manuscript.

Funding: This work was supported by the Consejo Nacional de Ciencia y Tecnología (CONACyT) of Mexico through a scholarship granted to Jean Pierre González-Gómez [No. 619431].

Institutional Review Board Statement: Not applicable.

Informed Consent Statement: Not applicable.

Data Availability Statement: The data used to support the findings of this study are available from the corresponding authors upon request.

Acknowledgments: The authors gratefully acknowledge CONACyT for the scholarship granted to Jean Pierre González-Gómez. As well, the authors would like to thank Hector Iván Hernández, Sergio Rivera, Melissa Ríos, and Célida Isabel Martínez-Rodríguez for their technical support.

Conflicts of Interest: The authors declare no conflict of interest.

References

1. Gutiérrez, D.; Rodríguez-Rubio, L.; Martínez, B.; Rodríguez, A.; García, P. Bacteriophages as weapons against bacterial biofilms in the food industry. *Front. Microbiol.* **2016**, *7*, 825. [CrossRef]
2. Myszka, K.; Czaczyk, K. Bacterial biofilms on food contact surfaces—A review. *Pol. J. Food Nutr. Sci.* **2011**, *61*, 173–180. [CrossRef]

3. Anand, S.; Singh, A. Resistance of the constitutive microflora of biofilms formed on whey reverse-osmosis membranes to individual cleaning steps of a typical clean-in-place protocol. *J. Dairy Sci.* **2013**, *96*, 6213–6222. [CrossRef] [PubMed]
4. Carter, M.Q.; Quinones, B.; He, X.; Zhong, W.; Louie, J.W.; Lee, B.G.; Yambao, J.C.; Mandrell, R.E.; Cooley, M.B. An environmental Shiga toxin-producing *Escherichia coli* O145 clonal population exhibits high-level phenotypic variation that includes virulence traits. *Appl. Environ. Microbiol.* **2016**, *82*, 1090–1101. [CrossRef]
5. Croxen, M.A.; Finlay, B.B. Molecular mechanisms of *Escherichia coli* pathogenicity. *Nat. Rev. Microbiol.* **2009**, *8*, 26–38. [CrossRef]
6. Orhan-Yanıkan, E.; da Silva-Janeiro, S.; Ruiz-Rico, M.; Jiménez-Belenguer, A.I.; Ayhan, K.; Barat, J.M. Essential oils compounds as antimicrobial and antibiofilm agents against strains present in the meat industry. *Food Control.* **2019**, *101*, 29–38. [CrossRef]
7. Patel, J.; Sharma, M.; Millner, P.; Calaway, T.; Singh, M. Inactivation of *Escherichia coli* O157:H7 attached to spinach harvester blade using bacteriophage. *Foodborne Pathog. Dis.* **2011**, *8*, 541–546. [CrossRef] [PubMed]
8. Sharma, M.; Ryu, J.-H.; Beuchat, L.R. Inactivation of *Escherichia coli* O157:H7 in biofilm on stainless steel by treatment with an alkaline cleaner and a bacteriophage. *J. Appl. Microbiol.* **2005**, *99*, 449–459. [CrossRef] [PubMed]
9. Polaska, M.; Sokolowska, B. Bacteriophages-a new hope or a huge problem in the food industry. *AIMS Microbiol.* **2019**, *5*, 324–346. [CrossRef]
10. Lang, L.H. FDA approves use of bacteriophages to be added to meat and poultry products. *Gastroenterology* **2006**, *131*, 1370. [CrossRef]
11. García, P.; Madera, C.; Martínez, B.; Rodríguez, A. Biocontrol of *Staphylococcus aureus* in curd manufacturing processes using bacteriophages. *Int. Dairy J.* **2007**, *17*, 1232–1239. [CrossRef]
12. Greer, G.G.; Dilts, B.D.; Ackermann, H.-W. Characterization of a *Leuconostoc gelidum* bacteriophage from pork. *Int. J. food Microbiol.* **2007**, *114*, 370–375. [CrossRef]
13. Heringa, S.D.; Kim, J.; Jiang, X.; Doyle, M.P.; Erickson, M.C. Use of a mixture of bacteriophages for biological control of *Salmonella enterica* strains in compost. *Appl. Environ. Microbiol.* **2010**, *76*, 5327–5332. [CrossRef]
14. Viazis, S.; Akhtar, M.; Feirtag, J.; Diez-Gonzalez, F. Reduction of *Escherichia coli* O157:H7 viability on hard surfaces by treatment with a bacteriophage mixture. *Int. J. Food Microbiol.* **2011**, *145*, 37–42. [CrossRef]
15. Ryan, E.M.; Alkawareek, M.Y.; Donnelly, R.F.; Gilmore, B.F. Synergistic phage-antibiotic combinations for the control of *Escherichia coli* biofilms in vitro. *FEMS Immunol. Med. Microbiol.* **2012**, *65*, 395–398. [CrossRef] [PubMed]
16. Simões, M.; Simões, L.E.; Vieria, M.J. A review of current and emergent biofilm control strategies. *LWT-Food Sci. Technol.* **2010**, *43*, 573–583. [CrossRef]
17. Sutherland, I.W.; Hughes, K.A.; Skillman, L.C.; Tait, K. The interaction of phage and biofilms. *FEMS Microbiol. Lett.* **2004**, *232*, 1–6. [CrossRef]
18. Mah, T.-F. Biofilm-specific antibiotic resistance. *Future Microbiol.* **2012**, *7*, 1061–1072. [CrossRef]
19. Onmaz, N.E.; Yildirim, Y.; Karadal, F.; Hizlisoy, H.; Al, S.; Gungor, C.; Disli, H.B.; Barel, M.; Dishan, A.; Akai Tegin, R.A.; et al. *Escherichia coli* O157 in fish: Prevalence, antimicrobial resistance, biofilm formation capacity, and molecular characterization. *LWT* **2020**, *133*, 109940. [CrossRef]
20. Bhardwaj, D.K.; Taneja, N.K.; Dp, S.; Chakotiya, A.; Patel, P.; Taneja, P.; Sachdev, D.; Gupta, S.; Sanal, M.G. Phenotypic and genotypic characterization of biofilm forming, antimicrobial resistant, pathogenic *Escherichia coli* isolated from Indian dairy and meat products. *Int. J. Food Microbiol.* **2021**, *336*, 108899. [CrossRef]
21. Montso, P.K.; Mlambo, V.; Ateba, C.N. Efficacy of novel phages for control of multi-drug resistant *Escherichia coli* O177 on artificially contaminated beef and their potential to disrupt biofilm formation. *Food Microbiol.* **2021**, *94*, 103647. [CrossRef] [PubMed]
22. Ribeiro, K.V.G.; Ribeiro, C.; Dias, R.S.; Cardoso, S.A.; de Paula, S.O.; Zanuncio, J.C.; de Oliveira, L.L. Bacteriophage Isolated from Sewage Eliminates and Prevents the Establishment of *Escherichia coli* Biofilm. *Adv. Pharm. Bull.* **2018**, *8*, 85–95. [CrossRef]
23. Beloin, C.; Roux, A.; Ghigo, J.M. Escherichia coli Biofilms. In *Bacterial Biofilms*; Romeo, T., Ed.; Springer: Berlin/Heidelberg, Germany, 2008; pp. 249–289.
24. Bridier, A.; Sanchez-Vizuete, P.; Guilbaud, M.; Piard, J.C.; Naitali, M.; Briandet, R. Biofilm-associated persistence of food-borne pathogens. *Food Microbiol.* **2015**, *45*, 167–178. [CrossRef]
25. Iniguez-Moreno, M.; Gutierrez-Lomeli, M.; Avila-Novoa, M.G. Kinetics of biofilm formation by pathogenic and spoilage microorganisms under conditions that mimic the poultry, meat, and egg processing industries. *Int. J. Food Microbiol.* **2019**, *303*, 32–41. [CrossRef]
26. Carson, L.; Gorman, S.P.; Gilmore, B.F. The use of lytic bacteriophages in the prevention and eradication of biofilms of *Proteus mirabilis* and *Escherichia coli*. *FEMS Immunol. Med. Microbiol.* **2010**, *59*, 447–455. [CrossRef]
27. Marshall, K.M.; Niebuhr, S.E.; Acuff, G.R.; Lucia, L.M.; Dickson, J.S. Identification of *Escherichia coli* O157:H7 meat processing indicators for fresh meat through comparison of the effects of selected antimicrobial interventions. *J. Food Prot.* **2005**, *68*, 2580–2586. [CrossRef]
28. Wang, C.; Hang, H.; Zhou, S.; Niu, Y.D.; Du, H.; Stanford, K.; McAllister, T.A. Bacteriophage biocontrol of Shiga toxigenic *Escherichia coli* (STEC) O145 biofilms on stainless steel reduces the contamination of beef. *Food Microbiol.* **2020**, *92*, 103572. [CrossRef] [PubMed]
29. Kang, J.-W.; Lee, H.-Y.; Kang, D.-H. Synergistic bactericidal effect of hot water with citric acid against *Escherichia coli* O157:H7 biofilm formed on stainless steel. *Food Microbiol.* **2021**, *95*, 103676. [CrossRef] [PubMed]

30. Iniguez-Moreno, M.; Avila-Novoa, M.G.; Gutierrez-Lomeli, M. Resistance of pathogenic and spoilage microorganisms to disinfectants in the presence of organic matter and their residual effect on stainless steel and polypropylene. *J. Glob. Antimicrob. Resist.* **2018**, *14*, 197–201. [CrossRef]
31. Ackermann, H.W.; DuBow, M.S. Practical applications of bacteriophages. In *Viruses of Prokaryotes: General Properties of Bacteriophages*; Ackermann, H.W., DuBow, M.S., Eds.; CRC Press: Boca Raton, FL, USA, 1987; pp. 143–158.
32. Adriaenssens, E.M.; Sullivan, M.B.; Knezevic, P.; van Zyl, L.J.; Sarkar, B.L.; Dutilh, B.E.; Alfenas-Zerbini, P.; Lobocka, M.; Tong, Y.; Brister, J.R.; et al. Taxonomy of prokaryotic viruses: 2018–2019 update from the ICTV Bacterial and Archaeal Viruses Subcommittee. *Arch. Virol.* **2020**, *165*, 1253–1260. [CrossRef] [PubMed]
33. Oechslin, F. Resistance Development to Bacteriophages Occurring during Bacteriophage Therapy. *Viruses* **2018**, *10*, 351. [CrossRef]
34. Chan, B.K.; Abedon, S.T.; Loc-Carrillo, C. Phage cocktails and the future of phage therapy. *Future Microbiol.* **2013**, *8*, 769–783. [CrossRef]
35. Tsonos, J.; Vandenheuvel, D.; Briers, Y.; De Greve, H.; Hernalsteens, J.-P.; Lavigne, R. Hurdles in bacteriophage therapy: Deconstructing the parameters. *Vet. Microbiol.* **2014**, *171*, 460–469. [CrossRef] [PubMed]
36. Sadekuzzaman, M.; Yang, S.; Mizan, M.F.R.; Ha, S.-D. Reduction of *Escherichia coli* O157:H7 in Biofilms Using Bacteriophage BPECO 19. *J. Food Sci.* **2017**, *82*, 1433–1442. [CrossRef]
37. Barbirz, S.; Becker, M.; Freiberg, A.; Seckler, R. Phage tailspike proteins with beta-solenoid fold as thermostable carbohydrate binding materials. *Macromol. Biosci.* **2009**, *9*, 169–173. [CrossRef]
38. Pires, D.P.; Oliveira, H.; Melo, L.D.R.; Sillankorva, S.; Azeredo, J. Bacteriophage-encoded depolymerases: Their diversity and biotechnological applications. *Appl. Microbiol. Biotechnol.* **2016**, *100*, 2141–2151. [CrossRef] [PubMed]
39. Mariana, N.S.; Salman, S.A.; Neela, V.; Zamberi, S. Evaluation of modified Congo red agar for detection of biofilm produced by clinical isolates of methicillin–resistance *Staphylococcus aureus*. *Afr. J. Microbiol. Res.* **2009**, *3*, 330–338.
40. Milanov, D.; Lazić, S.; Vidić, B.; Petrović, J.; Bugarski, D.; Šegulev, Z. Slime production and biofilm forming ability by *Staphylococcus aureus* bovine mastitis isolates. *Acta Vet.* **2010**, *60*, 217–226. [CrossRef]
41. Marques, S.C.; Rezende, J.d.G.O.S.; de Freitas Alves, L.A.; Silva, B.C.; Alves, E.; de Abreu, L.R.; Piccoli, R.H. Formation of biofilms by *Staphylococcus aureus* on stainless steel and glass surfaces and its resistance to some selected chemical sanitizers. *Braz. J. Microbiol.* **2007**, *38*, 538–543. [CrossRef]
42. Michu, E.; Cervinkova, D.; Babak, V.; Kyrova, K.; Jaglic, Z. Biofilm formation on stainless steel by *Staphylococcus epidermidis* in milk and influence of glucose and sodium chloride on the development of ica-mediated biofilm. *Int. Dairy J.* **2011**, *21*, 179–184. [CrossRef]
43. Alhede, M.; Qvortrup, K.; Liebrechts, R.; Hoiby, N.; Givskov, M.; Bjarnsholt, T. Combination of microscopic techniques reveals a comprehensive visual impression of biofilm structure and composition. *FEMS Immunol. Med. Microbiol.* **2012**, *65*, 335–342. [CrossRef] [PubMed]
44. Fratesi, S.E.; Lynch, F.L.; Kirkland, B.L.; Brown, L.R. Effects of SEM Preparation Techniques on the Appearance of Bacteria and Biofilms in the Carter Sandstone. *J. Sediment. Res.* **2004**, *74*, 858–867. [CrossRef]
45. Carey-Smith, G.V.; Billington, C.; Cornelius, A.J.; Hudson, J.A.; Heinemann, J.A. Isolation and characterization of bacteriophages infecting *Salmonella* spp. *FEMS Microbiol. Lett.* **2006**, *258*, 182–186. [CrossRef]
46. López-Cuevas, O.; Castro-del Campo, N.; León-Félix, J.; Chaidez, C. Characterization of bacteriophages with a lytic effect on various *Salmonella* serotypes and *Escherichia coli* O157:H7. *Can. J. Microbiol.* **2011**, *57*, 1042–1051. [CrossRef] [PubMed]
47. Rivera-Betancourt, M.; Shackelford, S.D.; Arthur, T.M.; Westmoreland, K.E.; Bellinger, G.; Rossman, M.; Reagan, J.O.; Koohmaraie, M. Prevalence of *Escherichia coli* O157:H7, *Listeria monocytogenes*, and *Salmonella* in two geographically distant commercial beef processing plants in the United States. *J. Food Prot.* **2004**, *67*, 295–302. [CrossRef] [PubMed]
48. Rossoni, E.M.M.; Gaylarde, C.C. Comparison of sodium hypochlorite and peracetic acid as sanitising agents for stainless steel food processing surfaces using epifluorescence microscopy. *Int. J. Food Microbiol.* **2000**, *61*, 81–85. [CrossRef]
49. Jamalludeen, N.; Johnson, R.P.; Friendship, R.; Kropinski, A.M.; Lingohr, E.J.; Gyles, C.L. Isolation and characterization of nine bacteriophages that lyse O149 enterotoxigenic *Escherichia coli*. *Vet. Microbiol.* **2007**, *124*, 47–57. [CrossRef]

MDPI

Article

Removal of Mixed-Species Biofilms Developed on Food Contact Surfaces with a Mixture of Enzymes and Chemical Agents

Maricarmen Iñiguez-Moreno [1,2], Melesio Gutiérrez-Lomelí [1,*] and María Guadalupe Avila-Novoa [1,*]

[1] Centro de Investigación en Biotecnología Microbiana y Alimentaria, Departamento de Ciencias Básicas, División de Desarrollo Biotecnológico, Centro Universitario de la Ciénega, Universidad de Guadalajara, Av. Universidad 1115, Col. Lindavista, 47820 Ocotlán, Jalisco, Mexico; mari.moreno2312@gmail.com

[2] Laboratorio Integral de Investigación en Alimentos, Tecnológico Nacional de México/Instituto Tecnológico de Tepic, Av. Tecnológico 2595, 63175 Tepic, Nayarit, Mexico

* Correspondence: melesio.gutierrez@academicos.udg.mx (M.G.-L.); avila.novoa@cuci.udg.mx (M.G.A.-N.)

Abstract: Sanicip Bio Control (SBC) is a novel product developed in Mexico for biofilms' removal. The aims of this study were to evaluate (i) the removal of mixed-species biofilms by enzymatic (protease and α-amylase, 180 MWU/g) and chemical treatments (30 mL/L SBC, and 200 mg/L peracetic acid, PAA) and (ii) their effectiveness against planktonic cells. Mixed-species biofilms were developed on stainless steel (SS) and polypropylene B (PP) in whole milk (WM), tryptic soy broth (TSB) with meat extract (TSB+ME), and TSB with chicken egg yolk (TSB+EY) to simulate the food processing environment. On SS, all biofilms were removed after treatments, except the enzymatic treatment that only reduced 1–2 $\log_{10}$ CFU/cm^2, whereas on PP, the reductions ranged between 0.59 and 5.21 $\log_{10}$ CFU/cm^2, being the biofilms developed in TSB+EY being resistant to the cleaning and disinfecting process. Higher reductions in microbial load on PP were reached using enzymes, SBC, and PAA. The employed planktonic cells were markedly more sensitive to PAA and SBC than were the sessile cells. In conclusion, biofilm removal from SS can be achieved with SBC, enzymes, or PAA. It is important to note that the biofilm removal was strongly affected by the food contact surfaces (FCSs) and surrounding media.

Keywords: stainless steel; polypropylene; organic matter; microbial resistance; peracetic acid

Citation: Iñiguez-Moreno, M.; Gutiérrez-Lomelí, M.; Avila-Novoa, M.G. Removal of Mixed-Species Biofilms Developed on Food Contact Surfaces with a Mixture of Enzymes and Chemical Agents. *Antibiotics* **2021**, *10*, 931. https://doi.org/10.3390/antibiotics10080931

Academic Editors: Luís Melo and Andreia S. Azevedo

Received: 9 July 2021
Accepted: 28 July 2021
Published: 30 July 2021

1. Introduction

Biofilms are growing communities of microorganisms adhered to a surface and were embedded in self-produced extracellular polymeric substances (EPS) [1]. The type and amount of EPS are strain-dependent and vary with the environmental conditions in which biofilms are formed. Nevertheless, the general composition of EPS includes polysaccharides, proteins, lipids, and extracellular DNA [1,2]. Biofilm development confers advantages to microbial cells, such as physical resistance to refrigeration, heat, desiccation, acidity, and salinity; mechanical resistance to liquid streams in pipelines; and chemical protection against antimicrobials and disinfectants [3,4]. Otherwise, biofilms cause corrosion in equipment, biofouling in water systems, and post-process contamination contributing to food spoilage. Recently, biofilms have been associated with the generation of foodborne diseases [5].

Almost 800 foodborne disease outbreaks are reported every year in the USA, causing approximately 15,000 foodborne illnesses, 800 hospitalizations, and 20 deaths [6]. The National Institutes of Health estimated that over 65% of microbial diseases are related to biofilm formation [7]. *Listeria monocytogenes*, *Salmonella*, and Shiga toxin-producing *Escherichia coli* are related to 82% of all hospitalizations and deaths in the USA. Another important biofilm-former microorganism commonly implicated in foodborne diseases is *Bacillus cereus* (2%) [6].

In the last decade, food industries have focused on managing food spoilage caused by biofilm-forming microorganisms, such as *Clostridium*, *Brochothrix thermosphacta*, Enterobacteriaceae, lactic acid bacteria, *Pseudomonas* spp., and *Bacillus* spp. [8,9]. The Food and Agricultural Organization of the United Nations [10] reported that over 25% of global food production is lost by microbial action (bacteria, yeasts, and molds). These microorganisms have been associated with biofilm development in dairy, meat, and egg processing industries [9].

Sanitization programs are the main alternative to biofilm control in the food industry [11], which comprise two phases: cleaning and disinfection [5]. The cleaning process removes food residues (proteins, fats, minerals deposits, sugars, and others) and 90% of the microorganisms from food contact surfaces (FCSs) [12]. Disinfection is the application of physical methods (UV light, cold plasma, ultrasound, etc.) or chemical agents (biocides and antimicrobials) to cause damage to or kill microorganisms [13]. Nonetheless, the use of physical [14] or chemical methods [15,16] is not enough to remove and eradicate the microorganisms within the biofilm, because the EPS occludes them with antimicrobial agents, reducing the shear forces [5,17]. Moreover, disinfectants cannot remove the biofilms; therefore, non-removed biofilm modify the surface charge and serve as a new substrate to other microorganisms, enabling them to restart the biofilm formation again [15]. Therefore, it is necessary to consider the introduction of different strategies, such as the use of enzymes, to achieve the biofilm removal [18]. Enzymes can kill bacteria and break down the biofilm structure due to EPS disruption [5,9]. CIP & GROUP is a Mexican company that recently developed Sanicip Bio Control (SBC). This is a cleaner and disinfectant product based on a mixture of high penetration surfactants, quaternary ammonium compounds of fifth-generation, and oxidant agents. The purpose of SBC is to achieve the biofilm removal from FCSs and kill the microorganisms that developed it.

Previous research demonstrated that *Salmonella*, *E. coli*, *L. monocytogenes*, *B. cereus*, and *Pseudomonas aeruginosa* developed mixed-species biofilms onto stainless steel (SS) and polypropylene B (PP), in whole milk (WM), and in culture media with egg yolk or meat extract [19]. In the food industry, biofilms are composed of multiple microorganisms; therefore, it is important to evaluate the cleaning and disinfection process in mixed-species biofilms developed under conditions that simulate food processing environments. Therefore, this study aimed to evaluate (i) the effect of enzymatic and chemical treatments on biofilm removal, (ii) examine the addition of peracetic acid to the disinfection process, and (iii) compare the effect of chemical agents against planktonic cells and mixed-species biofilms developed in the presence of organic matter or food residues.

2. Results

2.1. Microbicidal Activity against Planktonic Cells

E. coli, *Salmonella* Typhimurium, *Salmonella* Enteritidis, *P. aeruginosa*, *L. monocytogenes*, and *B. cereus* were reduced by 99.9999% (>6 $\log_{10}$ CFU/mL) after 30 s of contact with peracetic acid (PAA) or SBC, in suspensions without food residues. In the presence of food residues, SBC maintains its efficacy (99.9999%). However, the activity of PAA decreased in egg yolk and meat extract ($p \leq 0.05$) in comparison to that in WM. PAA reduced ~4 $\log_{10}$ CFU/mL of *E. coli*, *S.* Typhimurium, and *L. monocytogenes* in egg yolk; otherwise, in meat extract the reductions ranged from 0.86 to 2.30 $\log_{10}$ CFU/mL (Figure 1).

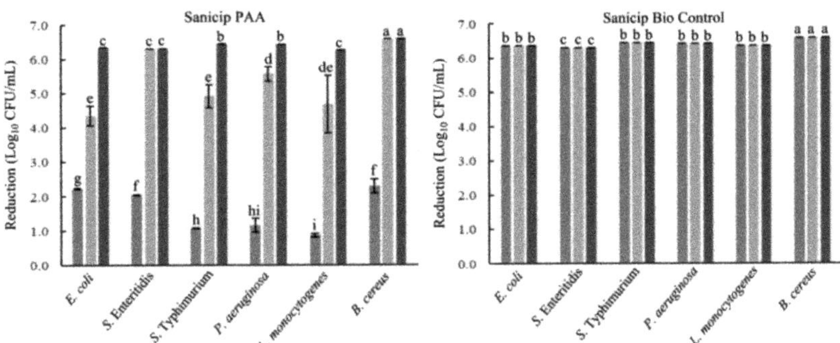

Figure 1. Microbicidal activity against planktonic cells in the presence of organic matter. Each bar represents the mean of three tests of the antimicrobial activity ± standard deviation. Sanicip PAA (200 mg/L) or Sanicip Bio Control (8 mL/L) in meat extract (■; 100 g/L); egg yolk (■; 100 mL/L); and whole milk (■; 100 mL/L). Bars within the same graph with different lower-case letter are significantly different according to Fisher's LSD test at $p \leq 0.05$.

2.2. *Biofilm Removal on SS*

To evaluate biofilm removal, mixed-species biofilms were developed on SS and PP coupons in three culture media. A mixture of proteolytic and amylolytic enzymes, and SBC were used for biofilm removal, and PAA was applied as a disinfectant (Figure 2). In biofilms developed on SS, the cellular densities ranged from 6.46 to 6.67 $\log_{10}$ (Table 1). Moreover, we analyzed the population before and after the treatments. In this regard, differences in the initial count between each species were observed ($p \leq 0.05$; Table 1). In addition, all treatments (except the enzymatic) reached over 6 $\log_{10}$ CFU/cm^2 of microorganism reduction in the mixed-species biofilms on SS. After enzymatic treatments, the counts of biofilms developed in tryptic soy broth (TSB) with 100 mL/L chicken egg yolk (TSB+EY) and whole milk (WM) were not different to those of their control ($p > 0.05$; Figure 3). However, the cell density of P. aeruginosa in biofilms developed in TSB+EY was reduced after the enzymatic treatment ($p \leq 0.05$). The same occurred with *E. coli* and *L. monocytogenes* in WM ($p \leq 0.05$, Table 1).

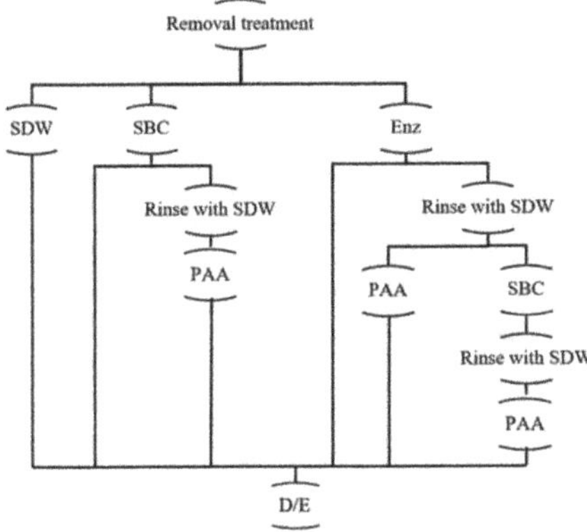

Figure 2. Treatments applied to removal and disinfection of mixed-species biofilms. SDW: sterile distilled water; SBC: Sanicip Bio Control (30 mL/L, 30 min, 25 °C); PAA: Sanicip PAA (200 mg/L, 10 min, 25 °C); Enz: enzymatic treatment (180 MWU/g, 30 min, 25 °C; MWU: modified Wohlgemuth unit); D/E: Dey/Engley broth (3 mL, 30 min).

Table 1. Microorganisms recovered from mixed-species biofilms developed on stainless steel AISI 304 before and after removal treatments.

Culture Media	Microorganism	Initial Count [a]	Treatments [b]					
			SDW	SBC	SBC + PAA	Enz	Enz + PAA	Enz + SBC + PAA
TSB + meat extract (100 g/L)	*E. coli*	4.41 ± 0.17 H[c]a[d]	4.70 ± 0.39 CDa	ND	ND	4.19 ± 0.22 BCDa	ND	ND
	S. Typhimurium	6.11 ± 0.13 BCa	5.28 ± 0.42 BCb	ND	ND	4.66 ± 0.28 Bc	ND	ND
	P. aeruginosa	6.26 ± 0.27 ABCa	4.83 ± 0.20 BCDb	ND	ND	4.32 ± 0.03 BCc	ND	ND
	L. monocytogenes	4.68 ± 0.10 GHa	4.41 ± 0.17 Da	ND	ND	3.43 ± 0.29 Fc	ND	ND
	B. cereus	ND	ND	ND	ND	ND	ND	ND
TSB + egg yolk (100 mL/L)	*E. coli*	5.61 ± 0.24 DEa	4.30 ± 0.65 Db	ND	ND	3.62 ± 0.54 DEFb	ND	ND
	S. Enteritidis	5.84 ± 0.13 CDa	5.28 ± 0.43 BCa	ND	ND	5.35 ± 0.40 Aa	ND	ND
	P. aeruginosa	6.52 ± 0.08 Aa	5.30 ± 0.39 BCb	ND	ND	3.86 ± 0.48 CDEFc	ND	ND
	L. monocytogenes	4.46 ± 0.28 Ha	4.31 ± 0.39 Dab	ND	ND	3.61 ± 0.38 EFb	ND	ND
	B. cereus	1.46 ± 0.28 Ia	ND	ND	ND	ND	ND	ND
Whole milk	*E. coli*	5.73 ± 0.38 DEa	5.39 ± 0.28 Ba	ND	ND	4.62 ± 0.32 Bb	ND	ND
	S. Typhimurium	6.50 ± 0.29 ABa	6.09 ± 0.28 Aab	ND	ND	5.91 ± 0.24 Ab	ND	ND
	P. aeruginosa	5.39 ± 0.21 EFa	4.90 ± 0.24 BCDb	ND	ND	4.09 ± 0.39 BCDEb	ND	ND
	L. monocytogenes	4.96 ± 0.38 FGa	5.03 ± 0.58 BCDa	ND	ND	3.49 ± 0.35 EFb	ND	ND
	B. cereus	ND	ND	ND	ND	ND	ND	ND

[a] Mean of three tests of initial population before removal treatments in $\log_{10}$ ± standard deviation (n = 3); [b] mean of three tests of microorganisms recovered after removal treatments in $\log_{10}$ ± standard deviation (n = 3); [c] values in the same column with different capital letter are significantly different ($p \leq 0.05$); [d] values in the same row with different lowercase letter are significantly different ($p \leq 0.05$).; initial cellular densities 6.46 ± 0.20, 6.67 ± 0.09 and 6.55 ± 0.27 Log_{10} CFU/cm^2 in TSB+ME, TSB+EY, and WM.; SDW: sterile distilled water; SBC: Sanicip Bio Control (30 g/L, 30 min, 25 °C); PAA: Sanicip PAA (200 mg/L, 10 min, 25 °C); Enz: enzymatic treatment (180 MWU/g, 30 min, 25 °C; MWU: modified Wohlgemuth unit). ND: not detected after removal treatment. Detection limit: 0.81 $\log_{10}$ CFU/cm^2.

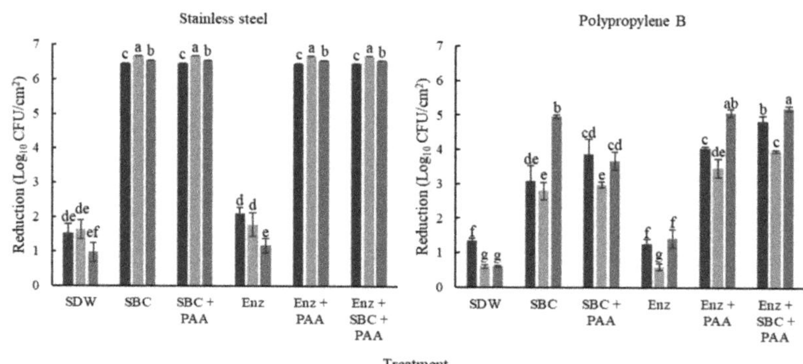

Figure 3. Reductions in mixed-species biofilms after removal and disinfection treatments. The biofilms were developed on stainless steel (SS) and polypropylene B (PP), in TSB with 100 g/L meat extract (■), TSB with 100 mL/L egg yolk (■), and whole milk (■) and were incubated at 25 °C for 240 h. Each bar represents the mean of three tests ± standard deviation of the means (n = 3) of cell density after removal treatments; SDW: sterile distilled water; SBC: Sanicip Bio Control (30 g/L, 30 min, 25 °C); PAA: Sanicip PAA (200 mg/L, 10 min, 25 °C); Enz: enzymatic treatment (180 MWU/g, 30 min, 25 °C; MWU: modified Wohlgemuth unit). Bars within the same graph with different lower-case letter are significantly different according to Fisher's LSD test at $p \leq 0.05$. Detection limits were 0.71 and 0.81 $\log_{10}$ CFU/cm^2 for PP and SS, respectively.

2.3. Biofilm Removal on PP

The initial biofilm counts were higher on PP (~7.49 $\log_{10}$ CFU/cm^2) than on SS ($p \leq 0.05$). Furthermore, the biofilms showed greater resistance on PP compared to on SS, with reductions between 0.59 and 5.21 $\log_{10}$ CFU/cm^2 ($p \leq 0.05$; Figure 3). However, significant differences were observed in the initial counts of each microorganism ($p \leq 0.05$). On PP, the low reductions were obtained with enzymatic treatments ($p > 0.05$, Figure 3). Moreover, E. coli and L. monocytogenes were recovered of biofilms developed in TSB with meat extract (TSB+ME) after treatments with SBC with or without the previous enzymes' application (Table 2).

Table 2. Microorganisms recovered from mixed-species biofilms developed on polypropylene before and after removal treatments.

Culture Media	Microorganism	Initial Count [a]	Treatments [b]					
			SDW	SBC	SBC + PAA	Enz	Enz + PAA	Enz + SBC + PAA
TSB + meat extract (100 g/L)	*E. coli*	5.30 ± 0.49 D[c]a[d]	4.87 ± 0.64 Fa	2.70 ± 0.39 DEb	ND	4.88 ± 0.36 DEFa	2.12 ± 0.13 Fc	ND
	S. Typhimurium	6.89 ± 0.38 CBa	6.42 ± 0.03 CDb	4.29 ± 0.21 Cc	3.33 ± 0.46 BCd	6.54 ± 0.17 ABb	3.52 ± 0.06 Bd	2.37 ± 0.16 CDe
	P. aeruginosa	7.11 ± 0.43 ABCa	5.62 ± 0.18 Eb	4.14 ± 0.63 CDc	2.89 ± 0.29 Cd	5.52 ± 0.31 Cb	2.79 ± 0.29 DEde	2.17 ± 0.41 DEe
	L. monocytogenes	5.42 ± 0.40 Da	4.93 ± 0.11 Fa	3.69 ± 0.32 BCb	ND	4.90 ± 0.33 DEa	3.01 ± 0.23 CDc	ND
	B. cereus	3.56 ± 0.01 Ea	2.08 ± 0.41 Gb	ND	ND	2.89 ± 0.42 Gb	ND	ND
TSB + egg yolk (100 mL/L)	*E. coli*	5.72 ± 0.41 Da	5.12 ± 0.32 Fa	ND	ND	4.47 ± 0.36 EFb	ND	ND
	S. Enteritidis	6.97 ± 0.27 ABCa	6.50 ± 0.06 Cb	4.07 ± 0.19 ABc	3.51 ± 0.29 Bd	6.42 ± 0.11 Bb	4.07 ± 0.10 Ac	3.05 ± 0.03 Be
	P. aeruginosa	6.63 ± 0.30 Ca	6.02 ± 0.17 DEb	3.48 ± 0.65 CDd	2.28 ± 0.21 Def	5.38 ± 0.17 Cc	2.45 ± 0.27 Eef	2.08 ± 0.25 DEf
	L. monocytogenes	5.03 ± 0.58 Ea	4.90 ± 0.11 Fa	ND	ND	4.51 ± 0.29 Fb	ND	ND
	B. cereus	3.06 ± 0.47 Ea	2.17 ± 0.22 Gb	ND	ND	ND	ND	ND
Whole milk	*E. coli*	7.33 ± 0.22 ABa	7.30 ± 0.45 Ba	2.20 ± 0.45 Ed	3.84 ± 0.40 ABc	6.18 ± 0.05 Bb	2.35 ± 0.21 DEFd	1.64 ± 0.32 Ee
	S. Typhimurium	8.11 ± 0.06 Aa	7.64 ± 0.27 Ab	3.13 ± 0.52 CDe	4.29 ± 0.47 Ad	6.92 ± 0.37 Ac	3.22 ± 0.15 Ce	3.10 ± 0.08 Ae
	P. aeruginosa	6.43 ± 0.17 Ca	5.62 ± 0.35 Eb	2.93 ± 0.23 CDd	3.73 ± 0.06 Bc	5.53 ± 0.27 Cb	2.84 ± 0.07 Dd	2.27 ± 0.11 BCd
	L. monocytogenes	5.70 ± 0.15 Da	5.20 ± 0.52 Eab	2.49 ± 0.11 DEc	ND	5.25 ± 0.03 CDb	1.94 ± 0.24 Gd	ND
	B. cereus	ND	ND	ND	ND	ND	ND	ND

[a] Mean of three tests of initial population before removal treatments in $\log_{10}$ ± standard deviation (n = 3); [b] mean of three tests of microorganisms recovered after removal treatments in log10 ± standard deviation (n = 3); [c] values in the same column with different capital letter are significantly different ($p \leq 0.05$); [d] values in the same row with different lowercase letter are significantly different ($p \leq 0.05$); initial cellular densities 7.18 ± 0.54, 7.07 ± 0.28, and 7.91 ± 0.07 log10 CFU/cm2 in TSB+ME, TSB+EY and WM; SDW: sterile distilled water; SBC: Sanicip Bio Control (30 g/L, 30 min, 25 °C); PAA: Sanicip PAA (200 mg/L, 10 min, 25 °C); Enz: enzymatic treatment (180 MWU/g, 30 min, 25 °C; MWU: modified Wohlgemuth unit); ND: not detected after removal treatment. Detection limit: 0.71 log10 CFU/cm^2.

Even when the initial counts of Salmonella and P. aeruginosa were similar ($p > 0.05$) in biofilms developed in TSB+ME and TSB+EY; Salmonella loads were higher than those of P. aeruginosa after all treatments ($p \leq 0.05$). Moreover, these microorganisms were recovered after all treatments applied on PP, with cellular densities between 2.37 and 4.63 Log_{10} CFU/cm^2.

In general, the microorganisms in the biofilms developed on PP were more resistant to removal and disinfection treatments than in the other residues or onto SS. L. monocytogenes were not recovered on biofilms developed in WM after treatments with PAA. Moreover,

B. cereus was not quantified before treatments in biofilms developed in WM, and in most cases, was fully reduced after the removal and disinfection process (Table 2).

2.4. Epifluorescent Microscopy and SEM Analyses

Representative micrographs of mixed-species biofilms developed on SS in the different culture media were obtained by SEM and epifluorescence microscopy (Figure 4). In concordance with the counting plate technique before removal treatments, metabolically active cells were observed by epifluorescent microscopy. Furthermore, EPS and food residues were observed (Figure 4A–C) and confirmed by SEM (Figure 4D–I). After the removal and disinfection process of biofilms developed on SS, metabolically active cells were not observed by epifluorescence microscopy, except on the coupons with enzymatic treatment. However, through SEM, some bacterial cells and residues of EPS were observed, particularly after treatments with enzymes (Figure 5). Otherwise, cells and EPS were detected on PP after all treatments. Nevertheless, the biofilms were considerably removed in comparison to the images obtained before the treatments (Figure 4). After enzymes use, the microorganisms were easily observed due to EPS removal (Figure 5).

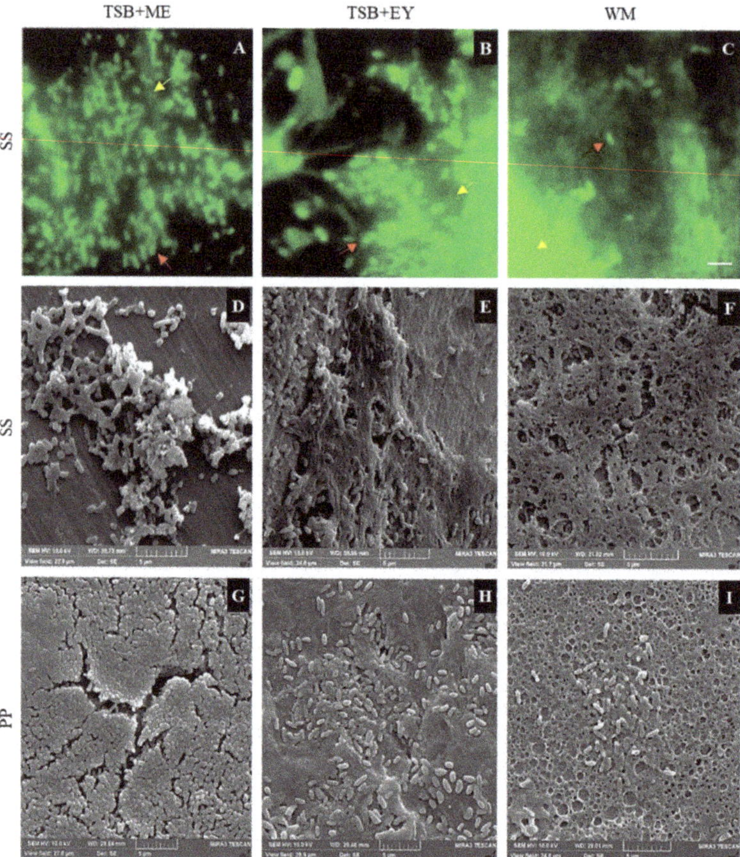

Figure 4. Micrographs of mixed-species biofilms. The micrographs were obtained by epifluorescence microscopy (**A–C**) and SEM (**D–I**) after 240 h of incubation at 25 °C of mixed-species biofilms in TSB with 100 g/L meat extract (TSB+ME), TSB with 100 mL/L egg yolk (TSB+EY), and whole milk (WM). The biofilms were developed on stainless-steel (SS) and polypropylene B (PP). The white bar scale indicates 5 μm. The red arrows shown the presence of metabolically active cells, whereas the yellow arrows indicate the presence of EPS and food residues.

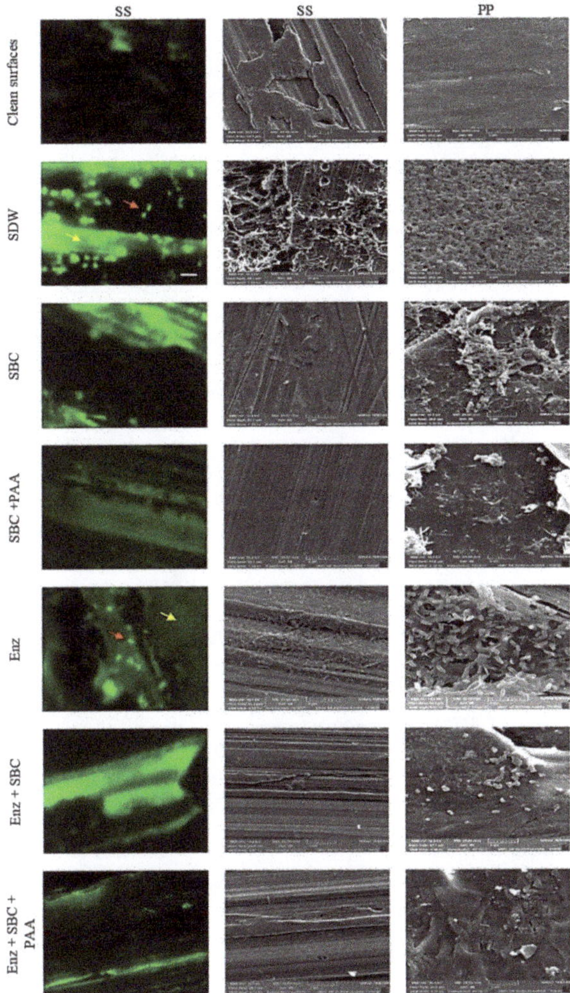

Figure 5. Micrographs of mixed-species biofilms after removal treatments. The mixed-species biofilms were developed on stainless steel (SS) and polypropylene B (PP) in whole milk during 240 h. The micrographs were obtained by epifluorescence microscopy and SEM after removal treatments SDW: sterile distilled water; SBC: Sanicip Bio Control (30 g/L, 30 min, 25 °C); PAA: Sanicip PAA (200 mg/L, 10 min, 25 °C); Enz: enzymatic treatment (180 MWU/g, 30 min, 25 °C; MWU: modified Wohlgemuth unit). The white bar scale indicates 5 μm. The red arrows shown the presence of metabolically active cells, whereas the yellow arrows indicate the presence of EPS and food residues.

3. Discussion

Biofilms contribute to pathogen spread and food contamination, cause damage to food processing equipment, and increase antimicrobial resistance, representing significant losses to the public and private sectors [5,20]. In this study, we assessed the effect of different treatments against planktonic and mixed-species biofilms developed under conditions that simulate a food processing environment. The assessed products reached reductions of >5 $\log_{10}$ CFU/mL against planktonic cells. A reduction of 5 $\log_{10}$ CFU/mL is the minimum to consider a disinfectant as effective [21,22]. An inappropriate cleaning process can leave up to 100 g/L of organic matter [23]; hence, microbicidal tests were also carried out in the presence of food residues. The efficacy of PAA was reduced in egg yolk and meat extract due to the fact that proteins and fats affect the availability of the oxidant agents [24]. Considering the time exposition used in this research (30 s), the reduction obtained (99.99%) was higher than that in other studies. For example, the products Suma Tab D4 and Suma Bac D10 (quaternary ammonium compounds, 240 and 740 mg/L, respectively) reduced 5 $\log_{10}$ of *L. monocytogenes* after 5 min in whole milk [23].

Biofilms are the main bacterial lifestyle in food processing environments, and sessile microorganisms are more resistant than are planktonic cells. Therefore, we evaluated the effect of different treatments on the removal of mixed-species biofilms developed in different culture media (tryptic soy broth (TSB) with chicken egg yolk (TSB+EY), and with meat extract (TSB+ME), and WM). SDW treatments (included as controls), showed significant reductions in *P. aeruginosa*, *S. enteritidis*, and *S. typhimurium*. SDW can cause cellular lysis, due to solutes absence, and it can also dissolve simple sugars, mineral salts [25], and some cellulose structures [26]. The use of chemical agents for biofilm control in food environments is not always effective; therefore, their efficacy should be improved by the combination of biological agents and physical methods.

Enzymes represent a great alternative for biofilm removal [27]. Biofilms in the food processing environment are composed of multiple microorganisms, resulting in an EPS with a heterogenic composition [28]. Recently, it has been demonstrated that polysaccharides in biofilms developed by Gram-negative bacteria, such as alginic acid, are the main component of the EPS matrix [13]. For example, the EPS matrix of S. Typhimurium is mainly composed of aggregative fimbriae and extracellular polysaccharides (cellulose) [29]. In contrast, proteins are the main compound in biofilms of Gram-positive bacteria [3]; however, they also produce polysaccharides as well as dextran [30]. Therefore, it is recommended to use a mixture of enzymes, because these molecules have specific activity [31]. In this study, the removal of mixed-species biofilms was evaluated using a mixture of alkaline protease and α-amylase. On SS surfaces, biofilm removal ranged between 93.4 and 96.3%. The low removal of biofilms developed in TSB+EY on PP (12.2%) was attributed to the high content of lipids in the egg yolk, which were not decomposed by the enzymes applied [32]. Ripolles-Avila et al. [3] achieved a removal of ~2.3 log CFU/cm^2 of S. Typhimurium on SS 304 with a mix of enzymes (protease, lipase, and amylase), which is in agreement with the findings in this study.

Kumari and Sarkar [33] used a serine protease, resulting in a complete reduction in *B. cereus* biofilms (4.08 $\log_{10}$ CFU/cm^2) developed in skim milk. The difference between this report and our results could be explained by the low cell density in the biofilms; moreover, after 24 h of incubation, the EPS matrix is not mature [19,34]. The EPS matrix is an important component of biofilms and represents more than 90% of the total mass of these structures [35]. The EPS matrix is the first resistance mechanism of the microorganisms in the biofilms against chemical and physical agents and environmental conditions. EPS components can react with the disinfectant molecules, protecting the microorganisms in the biofilm [36]. Recently, it has been reported that the application of an enzymatic cleaner (1 h at 50 °C) reduced 79.72% of *S. enterica* biofilm [28]. Nonetheless, to achieve this reduction, the samples were exposed for a least 1 h at 50 °C; this procedure is not viable for real conditions on an open surface in the food industry, a fact that was not considered in that report. Our study, however, was designed considering the application of the removal process on open surfaces in that environment.

To improve biofilm removal, enzymatic and chemical treatments were applied. With the combination of these treatments, microorganisms were not recovered from biofilms developed on SS. On PP, the reductions with the enzymatic and chemical treatments ranged from 3.06 to 4.76 $\log_{10}$ CFU/cm^2 (Figure 3), and these results were greater than those reported in other studies [36]. The selection of the type of FCS is vital in the food processing environment. The high biofilm removal on SS is related to its hydrophilic nature, the presence of metallic ions on the surface [37], the germicidal activity of the quaternary ammonium compounds, and the organic acids in the SCB. In previous research, it was demonstrated that PAA at 3500 mg/L kills the cells in biofilms of *Staphylococcus aureus* without removing them [15]. Moreover, lower reductions were obtained on PP, because hydrophobic surfaces increase cell aggregation and biofilm development [15,19]. The aqueous solution has minimal contact with the surface of PP; even SBC has quaternary ammonium compounds and surfactants that decrease the superficial tension of water, facilitating biofilm removal [27]. These compounds and the organic acids in their formulation promoted biofilm removal from SS.

Salmonella and *P. aeruginosa* were recovered in great amounts after the removal treatments. This is related to their high counts before the treatment; moreover, it has been demonstrated that the biofilm formation by *Salmonella* is favored in the presence of other bacteria such as *Pseudomonas* sp. and *Bacillus* sp. [38]. In addition, it was reported that *Salmonella* biofilms were more sensitive to disinfectants when they were developed on SS than on PP [15]. Microorganism aggregation within a three-dimensional structure can provide protection against biocides activity [39]. Almeida et al. [39] observed that two well-defined layers exist in tri-species biofilms, on the surface of *E. coli* and in the deep mixed regions of *L. monocytogenes* and *S. enterica*. This can explain the absence of *E. coli* after the application of all treatments.

Some studies have reported a higher resistance in sessile than in planktonic microorganisms to antimicrobials [40,41]. In line with this, the PAA reduced 2–3 $\log_{10}$ CFU/mL more in the assays with planktonic cells than in biofilms developed on PP treated with SBC or enzymes and then PAA. In addition to the EPS matrix, the presence of catalase in the microorganisms could play a role in peracetic acid decomposition. Unfortunately, the resistance mechanisms involved in mixed-species biofilms are not entirely clear [42].

Biofilms are complex structures composed of multilayers of microorganisms, EPS, and water channels [43]. The microorganisms in the biofilms are in different states: metabolically active, metabolically inactive, and dead cells. Therefore, it is important to use more than one technique for biofilm studies. Nowadays, it is difficult to use epifluorescence microscopy as a counting technique for cells in biofilms, because bacteria in a biofilm usually develop layers and residues such as TSB and EY emit strong auto-fluorescence [16,44]. However, it is possible to observe surviving cells after decontamination treatments, even in those treatments where it is not possible to achieve their expression in culture media, either by the detection limit of the technique or by the metabolic state of the bacteria (sub-lethally damaged cells or non-cultivable but metabolically active cells) [42]. SEM enables observing the architecture of the biofilm, without distinguishing living or dead cells [45]. Hence, for biofilm studies, complementary techniques should be used.

Disinfectant effectiveness on biofilms varies depending on disinfectant characteristics; type of surface; microorganisms in the biofilm; and other factors such as exposure time and temperature [15,41]. Furthermore, interspecies interactions generated within the biofilms have an effect on the dynamics and resistance within the biofilm [38]. Moreover, food residues rich in proteins, lipids, and carbohydrates decrease disinfectant effectiveness, thereby increasing bacteria survival and encouraging cross-contamination due to the increase in bacterial persistence on FCSs [5,38]. Currently, FCS coating, enzymatic disruption, quorum sensing inhibition, biosurfactants, bacteriophages, bacteriocins, essential oils, furanone derivates, high hydrostatic pressure, non-thermal plasma, ultrasound, and photocatalysis have been proposed for biofilm control [9,46]; however, these communities still represent a considerable challenge to food industries and scientists.

4. Materials and Methods

4.1. Bacterial Strains

The microorganisms used to biofilm formation were *E. coli* ATCC 11303, *S.* Typhimurium ATCC 14028, *S.* Enteritidis ATCC 13076, *P. aeruginosa* ATCC 15442, *L. monocytogenes* ATCC 19111, and *B. cereus* ATCC 14579 (vegetative stage). Before utilization, the microorganisms were incubated individually in TSB (Becton Dickinson Bioxon, Le Pont de

Claix, France) at 37 °C for 24 h in aerobic and static conditions to yield a final concentration of 10^7 CFU/mL.

4.2. Chemical and Enzymatic Agents

The assessed products were Sanicip Bio Control (active product obtained of the mixture of SBC 1 and SBC 2, National Sanitation Foundation (NSF) numbers 155919 and 155920, respectively) and Sanicip PAA (peracetic acid, PAA; 200 mg/L, NSF number 144381) (CIP & GROUP, Tlajomulco de Zuñiga, Mexico). Deterzyme 520/180 is a mixture of alkaline protease and α-amylase produced by *Bacillus licheniformis* and *Bacillus subtilis*, respectively (ENMEX, Tlalnepantla, Mexico), which was used to the assessment of biofilm removal.

4.3. Microbicidal Activity against Planktonic Cells

Bactericidal efficacy assays were performed according to AOAC Official Method 960.09 09 [21] with the products SBC (8 mL/L) and Sanicip PAA (200 mg/L). The concentrations used are approved for hard surfaces [23]. Briefly, 100 μL of overnight cultures (1×10^7 c/mL) were mixed by vortexing for 15 s (Vortex Genie 2, Model G-560), with 9.9 mL disinfectant solution with or without 100 g/L of organic matter (meat extract, Becton Dickinson & Co., Le Pont-de-Claix, France; egg yolk or WM processed at ultra-high temperatures purchased from a retail shop in Jalisco, Mexico). After 30 s, 100 μL of the assay mix was transferred to a new Eppendorf tube with 900 mL of Dey/Engley (D/E; Becton, Dickinson and Company, Le Pont de Claix, France) broth to neutralize the disinfectant activity. After 30 min of contact with D/E medium, the number of surviving bacteria was estimated by standard plate counting on tryptic soy agar (TSA; Becton Dickinson, Le Pont de Claix, France) and incubated at 37 °C for 24 h in aerobic conditions. Each assay was performed in triplicate. The percentage of reduction was calculated with the following formula:

$$Reduction\ (\%) = 100 - \frac{S(100)}{APC}$$

where S = surviving bacteria (CFU/mL) and ACP = aerobic counting plate initial (CFU/mL). The disinfectant was considered effective when it demonstrated a 99.999% bacterial reduction.

4.4. Biofilm Development

4.4.1. Contact Surfaces

The SS (AISI 304, 2 × 1 × 0.1 cm; CIMA Inoxidables, Guadalajara, Mexico) and PP coupons (2 × 1 × 0.2 cm; Plásticas Tarkus, Tlaquepaque, Mexico) were cleaned. Briefly, the surfaces were immersed in pure acetone (Fermont, Monterrey, Mexico) for 1 h to remove any debris and grease, immersed in neutral detergent (30 mL/L; Cip & Group S. de R.L., Tlajomulco de Zuñiga, Mexico) for 1 h, rinsed with sterile distilled water (SDW), cleaned with ethanol (700 mL/L; Hycel, Zapopan, Mexico), dried for 2 h at 60 °C, and sterilized by autoclaving (121 °C for 15 min) [47].

4.4.2. Biofilm Development and Quantification

Mixed-species biofilms were developed in three culture media: TSB with 100 mL/L chicken egg yolk, (TSB+EY), TSB with 100 g/L meat extract (TSB+ME), and WM. Briefly, each coupon was individually introduced into a new polypropylene tube (15 mL Centrifuge tube, Corning CentriStar, New York, NY, USA) containing 5 mL of the corresponding culture media and was inoculated with 25 μL of each bacterial species (1×10^6 CFU/mL). *S.* Typhimurium was used in biofilms developed in TSB+ME and WM and *S.* Enteritidis was inoculated in TSB+EY. The tubes with the coupons were incubated at 25 °C for 120 h. After this period, the coupons were removed from the tube, immersed into a new fresh medium and inoculated with the same microorganisms (1×10^6 CFU/mL), and incubated for 120 h. At the end of the incubation period, the coupons were removed from the tube using sterile forceps, rinsed by vortexing (150 rpm/10 s) in 5 mL of Dulbecco's phosphate-buffered saline (PBS; Sigma-Aldrich, St Louis, MO, USA). The conventional plate counting on tryptic soy agar with lactose (10 g/L; Sigma–Aldrich, St. Louis, MO, USA) and phenol red (0.1 g/L; Hycel, Zapopan, Jalisco, Mexico) was realized. For quantification of *E. coli* and *B. cereus* in multispecies biofilms, cefsulodin (50 μg/mL; Sigma–Aldrich, St. Louis, MO, USA) and polymyxin B (70 μg/mL; Sigma–Aldrich, St. Louis, MO, USA) were added to culture media, respectively. Petri dishes were incubated at 37 °C for 24 h. Colonies of *E. coli* and *L. monocytogenes* were yellow due to lactose fermentation; the other microorganism colonies were colorless. *Salmonella* and *P. aeruginosa* were distinguished using the oxidase test [19]. Each quantification was carried out in triplicate. Controls without microorganisms were included for the determination of contamination.

4.5. Removal and Disinfection Treatment Procedures

At the end of the incubation period, the coupons were removed from the tubes using sterile forceps and rinsed as above. Then, treatments to remove biofilms with SBC 30 mL/L or with a mixture of alkaline protease and α-amylase produced by *Bacillus licheniformis* and *Bacillus subtilis*, respectively (Deterzyme 520/80; ENMEX, Tlalnepantla, Mexico), were applied according to Figure 1. PAA at 200 mg/L was used as a disinfectant. Treatment with SDW at 30 °C/25 min was incorporated as a control. At the end of time exposure, each coupon was transferred to D/E broth. After 30 min in D/E broth, the surviving cells were estimated by counting plates, as in the Section 4.4.2. Each treatment was evaluated in triplicate.

4.6. Microscopy Analysis

4.6.1. Epifluorescent Microscopy

Before and after removal and disinfection treatments, biofilms developed on SS were rinsed with PBS as above, stained with 5(6)-carboxyfluorescein diacetate (CFDA, 10 μg/mL; Sigma Aldrich, St Louis, MO, USA), and dried in a cabinet biosafety level II. The CFDA excess was rinsed with SDW. The coupons were observed under a Nikon Eclipse E400 Epifluorescent Microscope using a 100× oil immersion lens and a BA 515 B-2A filter at 450–490 nm; at least 18 fields were observed [40].

4.6.2. Scanning Electron Microscopy (SEM) Analysis

Before and after removal and disinfection treatments, coupons of each material (SS and PP) were rinsed with PBS as above, then was immersed in 20 mL/L glutaraldehyde (DermoDex, Tlalpan, CDMX, Mexico) for 2 h at 4 °C to fix the biofilm. After serial dehydration in ethanol (30, 50, 60, 70, 90, and 95 mL/100 mL) for 10 min each at 4 °C, every coupon was rinsed (three 10 min rinses) in absolute ethanol [48]. Samples were dried and were coated with gold for 30 s [49]. Biofilms were observed under a TESCAN Mira 3 LMU Model field emission scanning electron microscope (Brno-Kohoutovice, Czech Republic).

4.7. Statistical Analysis

All of the experiments were performed in triplicate; the statistical analysis was carried out using ANOVA; the percentages data were arcsine square root transformed. The variances were examined by the least significant difference (LDS) test in the software Statgraphics Centurion XVI.I (Statpoint Technologies, Inc., Warrenton, VA, USA).

5. Conclusions

The resistance of mixed-species biofilms developed by *E. coli*, *S. typhimurium*, *S. enteritidis*, *P. aeruginosa*, *L. monocytogenes*, and *B. cereus* under conditions that simulate the dairy, meat, and egg processing industries was strongly affected by the type of FCS and surrounding media. The use of Sanicip Bio Control and enzymes plus Sanicip PAA were effective in removing the biofilms developed on SS. Hence, efforts should be conducted to prevent cell aggregation, promote the use of hydrophilic materials such as stainless steel, and use protocols of cleaning and disinfection based on the use of biological and chemical agents. Moreover, enzymatic agents are a great alternative to biofilm control in the food industry, establishing their use according to the type of food residues, potential microorganisms in the biofilm, and optimum temperature to maximize their activity. These results can contribute to applying novel approaches for controlling biofilms in food processing environments, improving food safety and quality.

Author Contributions: Conceptualization, methodology, validation, formal analysis, investigation, data curation, writing—original draft preparation and visualization, M.I.-M.; resources, writing—review and editing, supervision, project administration, and funding acquisition, M.G.-L. and M.G.A.-N. All authors have read and agreed to the published version of the manuscript.

Funding: This research was supported by CIP & GROUP S. de R.L and through a scholarship granted to Maricarmen Iñiguez-Moreno by the National Council of Science and Technology of Mexico [Grant number 404464].

Data Availability Statement: The data used to support the findings of this study are available from the corresponding authors upon request.

Acknowledgments: The authors gratefully acknowledge CIP & GROUP S. de R.L for the financial assistance, and particularly Ing. Jesús Casillas (General Director) and Ing. Sergio Miramontes (Technical Assessor), as well as the National Council of Science and Technology of Mexico, for the scholarship awarded to Maricarmen Iñiguez-Moreno.

Conflicts of Interest: The authors declare no conflict of interest. The funders had no role in the design of the study; in the collection, analyses, or interpretation of data; in the writing of the manuscript; or in the decision to publish the results.

References

1. Donlan, R.M. Biofilms: Microbial life on surfaces. *Emerg. Infect. Dis.* **2002**, *8*, 881–890. [CrossRef] [PubMed]
2. Flemming, H.C.; Wingender, J. The biofilm matrix. *Nat. Rev. Microbiol.* **2010**, *8*, 623–633. [CrossRef]
3. Ripolles-Avila, C.; García-Hernández, N.; Cervantes-Huamán, B.H.; Mazaheri, T.; Rodríguez-Jerez, J.J. Quantitative and compositional study of monospecies biofilms of spoilage microorganisms in the meat industry and their interaction in the development of multispecies biofilms. *Microorganisms* **2019**, *7*, 655. [CrossRef] [PubMed]
4. Flemming, H.-C.; Wingender, J.; Szewzyk, U.; Steinberg, P.; Rice, S.A.; Kjelleberg, S. Biofilms: An emergent form of bacterial life. *Nat. Rev. Microbiol.* **2016**, *14*, 563–575. [CrossRef] [PubMed]
5. Srey, S.; Jahid, I.K.; Ha, S.-D. Biofilm formation in food industries: A food safety concern. *Food Control* **2013**, *31*, 572–585. [CrossRef]
6. Mattia-Dewey, D.; Manikonda, K.; Hall, A.J.; Wise, M.E.; Crowe, S.J. Surveillance for foodborne disease outbreaks—United States, 2009–2015. *MMWR Surveill. Summ.* **2018**, *67*, 1–11. [CrossRef] [PubMed]
7. Jamal, M.; Ahmad, W.; Andleeb, S.; Jalil, F.; Imran, M.; Nawaz, M.A.; Hussain, T.; Ali, M.; Rafiq, M.; Kamil, M.A. Bacterial biofilm and associated infections. *J. Chin. Med. Assoc.* **2018**, *81*, 7–11. [CrossRef]
8. Majed, R.; Faille, C.; Kallassy, M.; Gohar, M. Bacillus cereus biofilms—Same, only different. *Front. Microbiol.* **2016**, *7*, 1–16. [CrossRef]
9. Motarjemi, Y.; Lelieveld, H. (Eds.) *Food Safety Management a Practical Guide for the Food Industry*, 14th ed.; Elsevier: San Diego, CA, USA, 2014; ISBN 9780123815040.
10. Gustavsson, J.; Cederberg, C.; Sonesson, U.; van Otterdijk, R.; Meybeck, A. *Global Food Losses and Food Waste—Extent, Causes and Prevention*; FAO: Rome, Italy, 2011.
11. Gibson, H.; Taylor, J.H.; Hall, K.E.; Holah, J.T. Effectiveness of cleaning techniques used in the food industry in terms of the removal of bacterial biofilms. *J. Appl. Microbiol.* **1999**, *87*, 41–48. [CrossRef]
12. Dat, N.D.; Manh, L.D.; Hamanaka, D.; Van Hung, D.; Tanaka, F.; Uchino, T. Surface conditioning of stainless steel coupons with skim milk, buttermilk, and butter serum solutions and its effect on bacterial adherence. *Food Control* **2014**, *42*, 94–100. [CrossRef]
13. Ripolles-Avila, C.; Ríos-Castillo, A.G.; Fontecha-Umaña, F.; Rodríguez-Jerez, J.J. Removal of *Salmonella enterica* serovar Typhimurium and *Cronobacter sakazakii* biofilms from food contact surfaces through enzymatic catalysis. *J. Food Saf.* **2019**, *40*, e12755. [CrossRef]
14. Lee, N.-Y.; Kim, S.-W.; Ha, S.-D. Synergistic effects of ultrasound and sodium hypochlorite (NaOCl) on reducing *Listeria monocytogenes* ATCC 19118 in broth, stainless steel, and iceberg lettuce. *Foodborne Pathog. Dis.* **2014**, *11*, 1–7. [CrossRef]
15. Iñiguez-Moreno, M.; Gutiérrez-Lomelí, M.; Guerrero-Medina, P.J.; Avila-Novoa, M.G. Biofilm formation by *Staphylococcus aureus* and *Salmonella* spp. under mono and dual-species conditions and their sensitivity to cetrimonium bromide, peracetic acid and sodium hypochlorite. *Braz. J. Microbiol.* **2018**, *49*, 310–319. [CrossRef]
16. Yang, Y.; Hoe, Y.W.; Zheng, Q.; Chung, H.; Yuk, H. Biofilm formation by *Salmonella* Enteritidis in a simulated liquid egg processing environment and its sensitivity to chlorine and hot water treatment. *Food Control* **2017**, *73*, 595–600. [CrossRef]

17. McDonnell, G.; Russell, A.D. Antiseptics and disinfectants: Activity, action, and resistance. *Clin. Microbiol. Rev.* **1999**, *12*, 147–179. [CrossRef] [PubMed]
18. Oulahal, N.; Martial-Gros, A.; Bonneau, M.; Blum, L.J. Removal of meat biofilms from surfaces by ultrasounds combined with enzymes and/or a chelating agent. *Innov. Food Sci. Emerg. Technol.* **2007**, *8*, 192–196. [CrossRef]
19. Iñiguez-Moreno, M.; Gutiérrez-Lomelí, M.; Avila-Novoa, M.G. Kinetics of biofilm formation by pathogenic and spoilage microorganisms under conditions that mimic the poultry, meat, and egg processing industries. *Int. J. Food Microbiol.* **2019**, *303*, 32–41. [CrossRef]
20. Zhao, X.; Zhao, F.; Wang, J.; Zhong, N. Biofilm formation and control strategies of foodborne pathogens: Food safety perspectives. *RSC Adv.* **2017**, *7*, 36670–36683. [CrossRef]
21. AOAC International Official Method. *Official Methods of Analysis of AOAC International*; Horwitz, W., Ed.; AOAC International Official Method: Arlington, VA, USA, 2005; pp. 1–3.
22. Food and Drugs Administration, Title 21—Food and Drugs Chapter I—Food and Drug Administration Department of Health and Human Services Subchapter B—Food for Human Consumption (Continued) Part 178—Indirect Food Additives: Adjuvants, Production Aids, and Sanitizers Subpart B–Substances Utilized To Control the Growth of Microorganisms. Available online: http://www.accessdata.fda.gov/scripts/cdrh/cfdocs/cfcfr/CFRSearch.cfm?fr=178.1010 (accessed on 3 December 2019).
23. Chaitiemwong, N.; Hazeleger, W.C.; Beumer, R.R. Inactivation of *Listeria monocytogenes* by disinfectants and bacteriophages in suspension and stainless steel carrier tests. *J. Food Prot.* **2014**, *77*, 2012–2020. [CrossRef] [PubMed]
24. Kalchayanand, N.; Koohmaraie, M.; Wheeler, T.L. Effect of exposure time and organic matter on efficacy of antimicrobial compounds against Shiga toxin-producing *Escherichia coli* and *Salmonella*. *J. Food Prot.* **2016**, *79*, 561–568. [CrossRef] [PubMed]
25. Parker, B.A. *JIFSAN Good Aquacultural Practices Program: Effective Cleaning and Sanitizing Procedures*; University of Maryland and the Johnson Diversey Corporation: College Park, ML, USA, 2007.
26. Burchard, W. Solubility and solution structure of cellulose derivatives. *Cellulose* **2003**, *10*, 213–225. [CrossRef]
27. Anand, S.; Singh, D.; Avadhanula, M.; Marka, S. Development and control of bacterial biofilms on dairy processing membranes. *Compr. Rev. Food Sci. Food Saf.* **2014**, *13*, 18–33. [CrossRef]
28. Nagraj, A.K.; Gokhale, D. Bacterial biofilm degradation using extracellular enzymes produced by *Penicillium janthinellum* EU2D-21 under submerged fermentation. *Adv. Microbiol.* **2018**, *8*, 687–698. [CrossRef]
29. De Rezende, C.E.; Anriany, Y.; Carr, L.E.; Joseph, S.W.; Weiner, R.M. Capsular polysaccharide surrounds smooth and rugose types of *Salmonella enterica* serovar Typhimurium DT104. *Appl. Environ. Microbiol.* **2005**, *71*, 7345–7351. [CrossRef] [PubMed]
30. Besrour-Aouam, N.; Fhoula, I.; Hernández-Alcántara, A.M.; Mohedano, M.L.; Najjari, A.; Prieto, A.; Ruas-Madiedo, P.; López, P.; Ouzari, H.-I. The role of dextran production in the metabolic context of *Leuconostoc* and *Weissella* Tunisian strains. *Carbohydr. Polym.* **2020**, *253*, 117254. [CrossRef] [PubMed]
31. Shen, H.B.; Chou, K.C. EzyPred: A top-down approach for predicting enzyme functional classes and subclasses. *Biochem. Biophys. Res. Commun.* **2007**, *364*, 53–59. [CrossRef]
32. De Reu, K.; Herman, L.; Heyndrickx, M.; De-waele, I. Risks of spoilage and *Salmonella* contamination of table eggs. *Lohmann Inf.* **2015**, *50*, 10–15.
33. Kumari, S.; Sarkar, P.K. Optimisation of *Bacillus cereus* biofilm removal in the dairy industry using an in vitro model of cleaning-in-place incorporating serine protease. *Int. J. Dairy Technol.* **2018**, *71*, 512–518. [CrossRef]
34. Bryers, J.D.; Ratner, B.D. Bioinspired implant materials befuddle bacteria. *ASM News* **2004**, *70*, 232–237.
35. Stoodley, P.; Sidhu, S.; Mather, M.; Boucek, A.; Hall-Stoodley, L.; Kathju, S. Kinetics and morphology of polymicrobial biofilm formation on polypropylene mesh. *FEMS Immunol. Med. Microbiol.* **2012**, *65*, 283–290. [CrossRef] [PubMed]
36. Molobela, I.P.; Cloete, T.E.; Beukes, M. Protease and amylase enzymes for biofilm removal and degradation of extracellular polymeric substances (EPS) produced by *Pseudomonas fluorescens* bacteria. *Afr. J. Microbiol. Res.* **2010**, *4*, 1515–1524.
37. Das, M.P. Effect of cell surface hydrophobicity in microbial biofilm formation. *Eur. J. Exp. Biol.* **2014**, *4*, 254–256.
38. Giaouris, E.; Heir, E.; Desvaux, M.; Hébraud, M.; Møretrø, T.; Langsrud, S.; Doulgeraki, A.; Nychas, G.-J.; Katcániová, M.; Czaczyk, K.; et al. Intra- and inter-species interactions within biofilms of important foodborne bacterial pathogens. *Front. Microbiol.* **2015**, *6*, 841. [CrossRef] [PubMed]
39. Almeida, C.; Azevedo, N.F.; Santos, S.; Keevil, C.W.; Vieira, M.J. Discriminating multi-species populations in biofilms with peptide nucleic acid fluorescence in situ hybridization (PNA FISH). *PLoS ONE* **2011**, *6*, e14786. [CrossRef]
40. Gorokhova, E.; Mattsson, L.; Sundström, A.M. A comparison of TO-PRO-1 iodide and 5-CFDA-AM staining methods for assessing viability of planktonic algae with epifluorescence microscopy. *J. Microbiol. Methods* **2012**, *89*, 216–221. [CrossRef] [PubMed]
41. Bridier, A.; Sanchez-Vizuete, P.; Guilbaud, M.; Piard, J.; Naïtali, M. Biofilm-associated persistence of food-borne pathogens. *Food Microbiol.* **2015**, *45*, 167–178. [CrossRef]
42. Bridier, A.; Sanchez-Vizuete, M.D.P.; Le Coq, D.; Aymerich, S.; Meylheuc, T.; Maillard, J.-Y.; Thomas, V.; Dubois-Brissonnet, F.; Briandet, R. Biofilms of a *Bacillus subtilis* hospital isolate protect *Staphylococcus aureus* from biocide action. *PLoS ONE* **2012**, *7*, e44506. [CrossRef] [PubMed]
43. Hobley, L.; Harkins, C.; MacPhee, C.E.; Stanley-Wall, N.R. Giving structure to the biofilm matrix: An overview of individual strategies and emerging common themes. *FEMS Microbiol. Rev.* **2015**, *39*, 649–669. [CrossRef]
44. Lee, K.; Millner, P.; Sharma, M.; Kim, M.S. Detection of bacterial biofilm on stainless steel by hyperspectral fluorescence imaging. In Proceedings of the Food Processing Automation Conference CD-Rom, Providence, RI, USA, 28–29 June 2008; pp. 1–6.

45. Alhede, M.; Qvortrup, K.; Liebrechts, R.; Høiby, N.; Givskov, M.; Bjarnsholt, T. Combination of microscopic techniques reveals a comprehensive visual impression of biofilm structure and composition. *FEMS Inmmunology Med. Microbiol.* **2012**, 1–8. [CrossRef]
46. Sharafutdinov, I.S.; Pavlova, A.S.; Khabibrakhmanova, A.M.; Faizova, R.G.; Kurbangalieva, A.R.; Tanaka, K.; Trizna, E.Y.; Baidamshina, D.R.; Bogachev, M.I.; Kayumov, A.R. Targeting *Bacillus cereus* cells: Increasing efficiency of antimicrobials by the bornyl-possessing 2(5H)-furanone derivative. *New Microbiol.* **2019**, *42*, 29–36. [PubMed]
47. Marques, S.C.; Rezende, J.G.O.S.; Alves, L.A.F.; Silva, B.C.; Alves, E.; Abreu, L.R.; Piccoli, R.H. Formation of biofilms by *Staphylococcus aureus* on stainless steel and glass surfaces and its resistance to some selected chemical sanitizers. *Braz. J. Microbiol.* **2007**, *38*, 538–543. [CrossRef]
48. Borucki, M.K.; Peppin, J.D.; White, D.; Loge, F.; Call, D.R. Variation in biofilm formation among strains of *Listeria monocytogenes*. *Appl. Environ. Microbiol.* **2003**, *69*, 7336–7342. [CrossRef] [PubMed]
49. Fratesi, S.E.; Lynch, F.L.; Kirkland, B.L.; Brown, L.R. Effects of SEM preparation techniques on the appearance of bacteria and biofilms in the carter sandstone. *J. Sediment. Res.* **2004**, *74*, 858–867. [CrossRef]

MDPI

Article

Thermal Shock and Ciprofloxacin Act Orthogonally on *Pseudomonas aeruginosa* Biofilms

Haydar Aljaafari [1,2], Yuejia Gu [1], Hannah Chicchelly [1] and Eric Nuxoll [1,*]

1 Department of Chemical and Biochemical Engineering, 4133 Seamans Center for the Engineering Arts and Sciences, University of Iowa, Iowa City, IA 52242, USA; haydar-aljaafari@uiowa.edu (H.A.); yuejia-gu@uiowa.edu (Y.G.); hannah.chicchelly@gmail.com (H.C.)
2 Department of Chemical Engineering, University of Technology, Baghdad 10066, Iraq
* Correspondence: eric-nuxoll@uiowa.edu; Tel.: +1-319-353-2377; Fax: +1-319-335-1415

Abstract: Bacterial biofilm infections are a major liability of medical implants, due to their resistance to both antibiotics and host immune response. Thermal shock can kill established biofilms, and some evidence suggests antibiotics may enhance this efficacy, despite having an insufficient effect themselves. The nature of this interaction is unclear, however, complicating efforts to integrate thermal shock into implant infection treatment. This study aimed to determine whether these treatments were truly synergistic or simply orthogonal (i.e., independent). *Pseudomonas aeruginosa* biofilms of different architectures and stationary-phase population density were subjected to various thermal shocks, antibiotic exposures, or combinations thereof, and examined either immediately after treatment or after subsequent reincubation. Population decreases from the combination treatment matched the product of the decreases of individual treatments, indicating their orthogonality. However, reincubation showed binary behavior, where biofilms with an immediate population decrease beyond a critical factor (~10^4) died off completely during reincubation, while biofilms with a smaller immediate decrease regrew. This critical factor was independent of the initial population density and the combination of treatments that achieved the immediate decrease. While antibiotics do not appear to enhance thermal shock directly, their contribution to achieving a critical population decrease for biofilm elimination can make the treatments appear strongly synergistic, strongly decreasing the intensity of thermal shock needed.

Keywords: biofilms; prosthesis-related infections; heat shock; ciprofloxacin; antibacterial agents

Citation: Aljaafari, H.; Gu, Y.; Chicchelly, H.; Nuxoll, E. Thermal Shock and Ciprofloxacin Act Orthogonally on *Pseudomonas aeruginosa* Biofilms. *Antibiotics* **2021**, *10*, 1017. https://doi.org/10.3390/antibiotics10081017

Academic Editors: Luís Melo and Andreia S. Azevedo

Received: 20 June 2021
Accepted: 16 August 2021
Published: 21 August 2021

1. Introduction

More than 750,000 knee replacement and 500,000 hip replacement surgeries are performed each year in the United States [1], and these numbers are expected to increase exponentially in the next decade [2]. In total, 1% to 4% of the knee replacement and 1% to 2% of the hip replacement procedures are followed by incidences of periprosthetic joint infection [3–5]. The pathogens in these infections typically organize themselves in a densely populated thin layer of polysaccharides, proteins, and DNA called a biofilm, in which they exhibit a phenotype 20–100 times more resistant to antibiotics and host immune response than their planktonic phenotype [6–10]. Thus, the current standard of care is high doses of antibiotics and surgical explantation of the implant with its surrounding infected tissue, followed eventually by implantation of a replacement device [11–13]. Though this is successful in over 90% of cases [14,15], the new implant has a higher risk of infection than the original one [16]. These multiple invasive procedures expose the patient to physical risk and low quality of life, in addition to significant financial costs [11,17]. The incidence of infection has persisted despite decades of effort to create surfaces that prevent biofilm formation [18–24] and to develop methods to eradicate established biofilms [25–32], none of which have progressed to clinical implementation.

Thermal shock has been demonstrated as a means of deactivating established biofilms but may also damage adjacent tissue [33,34]. Recent studies have suggested a synergism between antibiotics and thermal shock, with the combined treatment decreasing biofilm population density more than either treatment alone, or even by the product of their individual effects [35–38]. The nature of this interaction is poorly understood. One hypothesis is that biofilms may have a critical population density below which they become non-viable, and that any combination of approaches that drops the population below that level will result in complete elimination of the biofilm, even if the individual approaches are not nearly so effective. This study investigated this hypothesis in *Pseudomonas aeruginosa* biofilms. Using biofilms of significantly different architecture and initial population density, it demonstrates not a critical population level but rather a critical population decrease, beyond which the population proceeds to zero.

Two different protocols were used to culture *P. aeruginosa* biofilms with significantly different population density and architecture: Shaker table (ST) and drip flow reactor (DFR). Biofilms of each type were subjected to thermal shocks ranging from 50 °C for 1 min to 70 °C for 30 min, with or without 4 h of prior exposure to ciprofloxacin (CP) concentrations ranging from 0.25 to 64 μg/mL. Shocked biofilms were either immediately enumerated or reincubated for 1 or 2 days before enumeration. Biofilms of each type were similarly exposed to antibiotics and re-incubation without thermal shock.

2. Results

2.1. Population Density, Architecture, and Thermal Susceptibility

Biofilms grown using DFR and ST protocols had markedly different population densities (as measured in colony forming units (CFU) per cm^2), spanning from sparsely populated ($10^{7.13\pm0.58}$ CFU/cm^2) biofilms with individual micron-scale features ~50 μm in height (ST) to densely populated ($10^{8.3\pm0.4}$ CFU/cm^2) carpets over 100 μm thick (DFR). Figure 1 demonstrates these architectural differences with confocal fluorescent microscopy images of ST (panel a) and DFR (panel b) biofilms, as well as comparing their population densities (panel c).

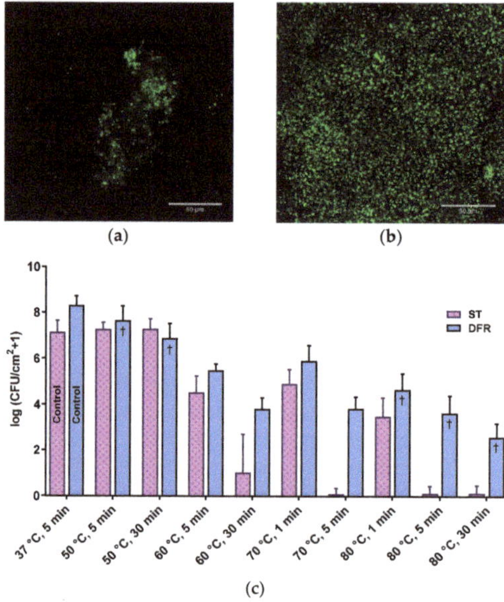

Figure 1. Culture protocols produce biofilm with starkly different population density, architecture, and thermal susceptibility. (**a**,**b**) are overhead views of confocal fluorescent images of shaker table (**a**) and drip flow reactor (**b**) biofilms. (**c**) Effect of thermal shock on population density of *P. aeruginosa* ST and DFR biofilms. † from [35].

The thermal susceptibility of the biofilms was similarly different at modest temperatures (50 °C), with DFR biofilms decreasing by up to 1.5 orders of magnitude while ST biofilms showed no effect. At high temperatures (80 °C), however, the thermal susceptibility of the different biofilms converges, with both biofilm types decreasing by 3.5 orders of magnitude after 1 min of exposure. For longer exposures, ST biofilm populations appear to drop off completely.

2.2. Re-Incubation

Re-incubation of thermally shocked DFR biofilms showed two opposing trends. After shocks sufficient to drop the population density below $10^{4.5}$ CFU/cm^2, the population density continued to decrease during re-incubation, dying off completely within a few hours. After milder thermal shocks, however, the biofilms regrew during reincubation, achieving their previous stationary-phase population density within a day. This is shown graphically in Figure 2 for eight thermal shock conditions, two of which typically dropped DFR biofilms to ~$10^{4.5}$ CFU/cm^2. At those two conditions, both trends can be observed, with 60% (6 of 10) of the biofilms shocked at 60 °C for 15 min dying off while the remainder grew back (Figure 2a), and 66% (8 of 12) of the biofilms shocked at 70 °C for 3 min dying off while the remainder grew back (Figure 2b).

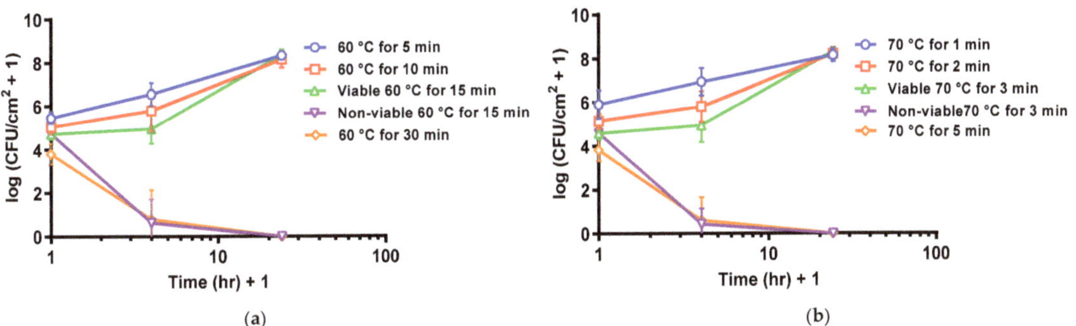

Figure 2. Critical population density for DFR biofilm reincubation. (**a**) Population density for DFR biofilm reincubation after thermal shock at 60 °C. (**b**) Population density for DFR biofilm reincubation after thermal shock at 70 °C. Error bars indicate standard deviation for at least six slides from three different dishes.

2.3. Antibiotic Exposure

Figure 3 shows that the relationship between the *P. aeruginosa* population density and CP concentration follows a power law for both DFR and ST biofilms, albeit shifted, with the same dosage having less efficacy on DFR biofilms than on ST ones. Panel (a) shows that for ST biofilms, four hours of exposure failed to decrease the population density by four orders of magnitude even at concentrations over 60 times higher than physiological dosing 4 µg/mL [39]. Increasing the exposure time to 24 h increased efficacy but still did not eliminate the biofilm, and with further exposure (48 h) the population density actually recovered rather than further decreased. The effect of CP is much smaller on DFR biofilms (Panel b) with only modest decreases in population density even at grossly toxic concentration (64 µg/mL) and long exposure times, though under these circumstances, the population density is at least still trending downward with exposure time.

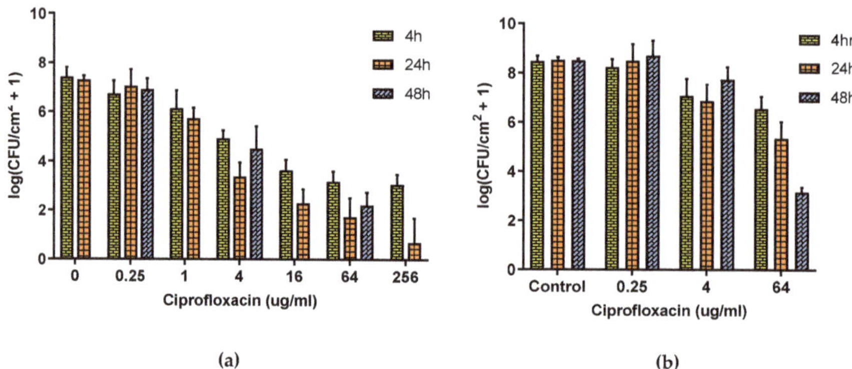

Figure 3. Effect of Ciprofloxacin on ST and DFR biofilms. (**a**) Effect of Ciprofloxacin on ST biofilm. (**b**) Effect of Ciprofloxacin on DFR biofilm. Error bars indicate standard deviation for at least six slides from three different dishes.

2.4. Combined Antibiotic and Thermal Exposure of Shaker Table (ST) Biofilms

Five different thermal shock protocols (50 °C for 5 or 30 min, 60 °C for 1 or 5 min, 70 °C for 1 min) were applied to ST biofilms after 4 h of CP (0.25 or 4.0 µg/mL) exposure at 37 °C. For each protocol and CP concentration, three reincubation conditions were investigated: 0, 24, or 48 h in fresh media at the designated CP concentration. The results for each protocol are shown in their own panel of Figure 4, alongside controls with no thermal shock. These controls include: No treatment (first, or leftmost, bar of each panel), with an average pre-treatment population density of $10^{7.13}$ CFU/cm^2; 0.25 µg/mL CP exposure for 4 or 24 h (3rd and 4th bars, respectively), showing that this exposure by itself had no significant effect on population density; and 4 µg/mL CP exposure for 4, 24, or 48 h (7th, 8th, and 9th bars, respectively), showing that, by itself, this maximum physiological dose of ciprofloxacin only reduced the population density to $10^{4.92 \pm 0.33}$ CFU/cm^2 after 4 h of exposure and $10^{3.37 \pm 0.58}$ after 24 h of exposure, recovering back to $10^{4.51 \pm 0.9}$ after 48 h of exposure.

The second bar of each chart shows the effect of the thermal shock by itself (i.e., no antibiotic exposure), with only the shocks at 60 °C for 5 min (Figure 4d) or 70 °C for 1 min (Figure 4e) showing any significant effect. Combining thermal shock with exposure to 0.25 µg/mL CP, thermal shocks that had no effect by themselves also had no effect when added with 0.25 µg/mL antibiotic. There is a significant effect for the 50 °C for 30 min (Figure 4b) shock after 4 h of antibiotic exposure, but this disappeared with 24 h of exposure and may be an anomaly. For thermal shocks that did have a significant effect by themselves, 4 h of exposure to 0.25 µg/mL CP may have increased the effect, albeit not significantly. Further exposure for 24 h has conflicting results, with the enhancement becoming significant with a 60 °C shock for 5 min (Figure 4d), while allowing regrowth after 70 °C shocks for 1 min (Figure 4e). No consistent synergism (or antagonism) is seen.

Combining thermal shock with exposure to 4.0 µg/mL CP, shocks at 50 °C for 5 min (Figure 4a) and 60 °C for 1 min (Figure 4c) appear to have no effect; the results are essentially the same with or without the shock. Since these shocks also had no effect by themselves, one can state that the effects of the two treatments are additive, with the log reduction of the combined treatment matching the sum of the log reductions for each treatment by itself. Similarly, for more aggressive shocks (60 °C for 5 min, or 70 °C for 1 min), both the thermal shock and the 4 h exposure to 4 µg/mL CP have an effect by themselves, and the log reduction in population density for the combined treatment roughly matches the sum of the log reductions for each individual treatment. Only for the intermediate shock at 50 °C for 30 min (Figure 5b) does the combined treatment prompt a log reduction larger than the sum of the individual treatments, suggesting a synergism between the treatments.

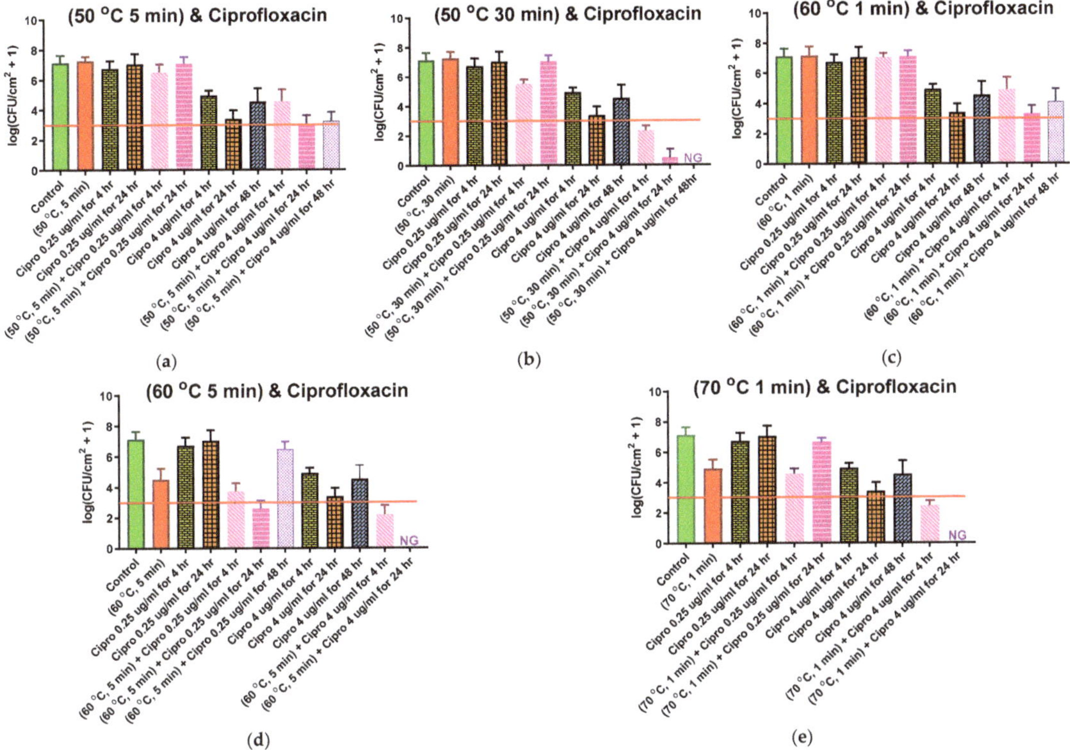

Figure 4. Combined ciprofloxacin and thermal shock effect on the *P. aeruginosa* shaker table biofilm population. Each panel shows results for the thermal shock and antibiotic exposure indicated. Red horizontal lines show the critical population density below which thermal shocked bacterial biofilms are not viable. Error bars indicate standard deviation for at least six slides from three different dishes.

Additional exposure to 4 µg/mL CP after these more aggressive thermal shocks has a dramatically different result, however. After 24 h of exposure, the biofilms shocked at 60 °C for 5 min or 70 °C for 1 min have no detectable CFU, see Figure 4d,e respectively. After 48 h, the biofilm shocked at 50 °C for 30 min (Figure 4b) also has no detectable CFU. These treatments appear highly synergistic.

2.5. Combined Antibiotic and Thermal Exposure of Drip Flow Reactor (DFR) Biofilms

Four different thermal shock protocols (60 °C for 5 or 10 min, 70 °C for 1 or 2 min) were applied to DFR biofilms in combination with subsequent CP (0.25 or 4.0 µg/mL) exposure at 37 °C for 4, 24, or 48 h. These protocols were chosen because they each cause a large population density reduction in DFR biofilms but not so large that the biofilm eventually dies off, as seen in Figure 2. The results for each protocol are shown in their own panel of Figure 5, alongside controls with no thermal shock. These controls include: No treatment (first, or leftmost, bar of each panel), with an average pre-treatment population density of $10^{8.3}$ CFU/cm^2; 0.25 µg/mL CP exposure for 4 or 24 h (3rd and 4th bars, respectively), showing that this exposure by itself had no significant effect on population density; and 4 µg/mL CP exposure for 4, 24, or 48 h (7th, 8th, and 9th bars, respectively), showing that, by itself, this maximum physiological dose of CP only reduced the population density by an average log reduction of about 1.5 after 4 or 24 h of exposure, and that biofilms actually recovered to approximately their pre-treatment population density within 48 h, despite continuous antibiotics exposure.

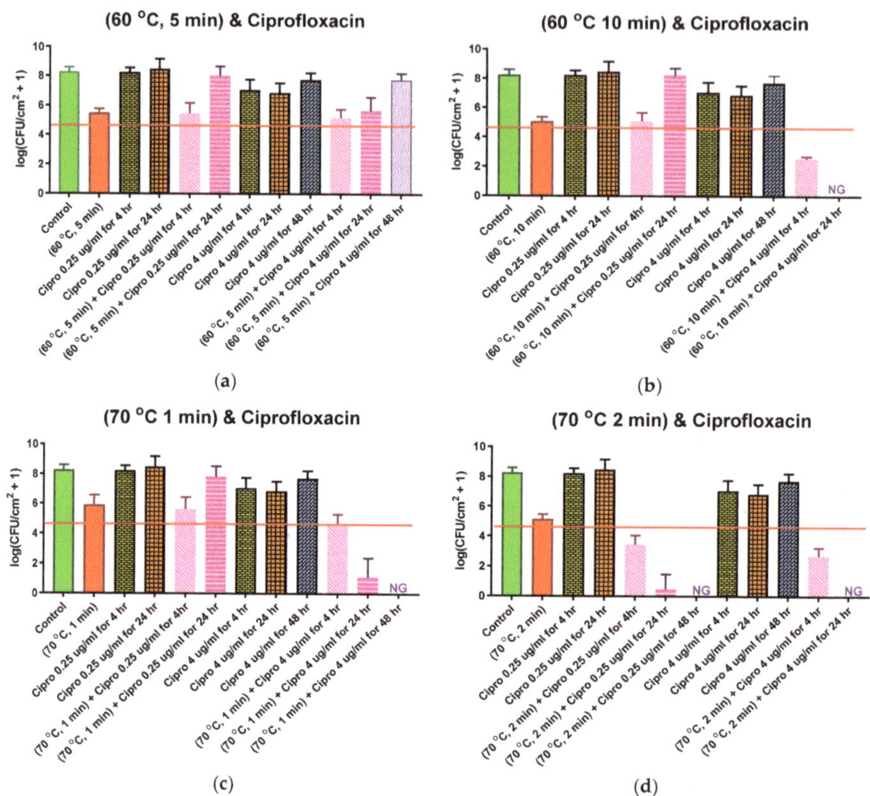

Figure 5. Combined ciprofloxacin and thermal shock effect on the *P. aeruginosa* drip flow reactor biofilm population. Each panel shows results for the thermal shock and antibiotic exposure indicated. Red horizontal lines show the critical population density below which thermal shocked bacterial biofilms are not viable. Error bars indicate standard deviation for at least six slides from three different dishes.

The second bar of each chart shows the effect of the thermal shock by itself (i.e., no antibiotic exposure), with log reductions of 2.3–3.4. Adding four hours of exposure to 0.25 μg/mL CP, the thermal shock had virtually no additional effect except at the highest temperature for its longest shock (70 °C for 2 min), where an additional log reduction of 1.7, to $10^{3.5}$ CFU/mL, was observed (Figure 5d). Increasing the antibiotic exposure to 24 h simply allowed the biofilm to grow back to its pre-treatment population density, except in the latter case, where the bacteria continued to die off, with no detectable CFU after 48 h.

Adding four hours of exposure to 4.0 μg/mL CP after the thermal shock, however, reduced the population density beyond the thermal shock alone in each case. For the shorter shocks at each temperature (Figure 5a,c), the log reductions of the combined treatments were slightly less than the sum of the log reductions of each treatment alone, see Figure 5b,d. The longer shocks, which had larger log reductions by themselves, also had larger log reductions from the combined treatment, even larger than the sum of the individual treatments. With longer antibiotic exposure, the biofilm population plummeted for most cases, with no detectable CFU after 24 h for the longer shocks at each temperature, and no detectable CFU after 48 h for the 1 min shock at 70 °C. Only for the 5 min shock at 60 C°, which had the highest population density after 4 h of 4 μg/mL CP exposure, did the biofilm regrow with continued antibiotic exposure, eventually (48 h) approaching its pre-treatment population density. Like the ST biofilms, these results suggest significant synergism between thermal shock and antibiotic exposure.

3. Discussion

Biofilm infections are particularly problematic because they cannot be eliminated by our primary in vivo treatment against bacteria, antibiotics. Ciprofloxacin's ability to eliminate this strain of bacteria in the planktonic phenotype has been previously reported [35]. In the biofilm phenotype, however, dramatically higher drug concentrations are needed for comparable population reduction. While this reduction follows a power-law relationship with concentration, it levels out without elimination of the biofilm, as reported previously for MBEC assay biofilms [35] and indicated here in Figure 3 for ST biofilms, where at grossly toxic antibiotic concentrations, the population density appears to asymptote at ~10^3 CFU/cm^2 after 4 h. The population density decreases further over the next 20 h but recovers again over the following day. For DFR biofilms, the decrease is irrelevant even at grossly toxic concentrations. Other approaches are clearly needed for biofilm infection mitigation.

Thermal shock is known to kill biofilms, but implementation in vivo requires minimizing the severity of the shock in order to minimize the accompanying damage to the adjacent tissue. Previous reports have indicated that antibiotics increase the efficacy of thermal shock, mitigating biofilms with less aggressive thermal treatment [35]. In some cases, the two approaches appear to be synergistic, with the reduction from the combined treatment exceeding the sum of the reductions from the individual treatments. The nature of this synergism was unknown, so this project aimed to investigate it, specifically scrutinizing the effect of re-incubation, with or without antibiotics, for biofilms spanning a wide range of architecture and population density.

As discussed earlier, the ST and DFR growth protocols involve starkly different growing conditions, which provide biofilms with dramatically different population density, structure, and maturity, as illustrated in Figure 1. These differences also prompt different thermal susceptibilities, with DFR biofilms strongly impacted by 50 °C exposure while ST biofilms are unaffected by it. At higher temperatures, however, their susceptibility (as measured by immediate population decrease) converges.

Immediate population decrease does not appear to completely quantify thermal susceptibility, though. After the thermal shock is removed and incubation conditions are restored, some DFR biofilms will recover to their pre-treatment population density, while others will continue to die off until no CFU are detectable. Figure 2 shows that this behavior can be predicted from the population density immediately after the thermal shock. DFR biofilms with a population density above ~$10^{4.5}$ CFU/cm^2 grow back, while biofilms with a population density below that value die off, regardless of the temperature of the shock. Similar behavior was previously reported for ST biofilms, but in that case, the critical population density was ~10^3 CFU/cm^2 [40]. Notably, this difference in critical population density (by a factor of ~$10^{1.5}$) is the same as the difference in the initial population densities of the two biofilms. In both cases, when the population density dropped by over four orders of magnitude, the biofilm died off, while smaller population density reductions resulted in the biofilm growing back completely, no matter what temperature and exposure time were used to achieve the reduction. These results suggest that mitigating biofilms is not based on driving its population density below a particular critical quorum, but rather achieving a particular population density reduction. This is encouraging since, at higher temperatures, this reduction appears to be the same regardless of the type of biofilm or its initial population density, as shown in Figure 1.

Figures 4 and 5 suggest that this re-incubation behavior is responsible for the perceived synergism between thermal shock and antibiotics. Each panel in Figure 5 includes a horizontal line at the critical population density identified in Figure 2. Similarly, each panel in Figure 4 includes a horizontal line at the critical population density previously reported for ST biofilms [40]. Looking at combination treatments that resulted in populations above the critical value after 4 h, the decrease in population density for the combined thermal shock + antibiotics treatment is not significantly different than the sum of the decreases by thermal shock alone and antibiotics alone. The only exception to this in either figure is for ST biofilms exposed to 0.25 μg/mL ciprofloxacin and shocked at 50 °C for 30 min. In every other case, the treatments appear to be simply additive, indicating that their mechanisms are orthogonal and non-overlapping but not synergistic. Even for the 50 °C/30 min/0.25 μg/mL case, the enhanced decrease disappears within 24 h.

On the other hand, when the population reduction of the combined thermal shock + antibiotic treatment brings the population density below the critical value, this reduction is significantly larger than the sum of the reductions of its individual components, and this difference becomes pronounced at longer times as the population of CFU drops to zero. This is demonstrated five times in Figure 4 (50 °C for 30 min with 4 μg/mL; 60 °C for 5 min with 0.25 μg/mL) and Figure 5 (60 °C for 10 min with 4 μg/mL; 70 °C for 1 min with 4 μg/mL; 70 °C for 2 min with 0.25 μg/mL) with a wide range of temperature/time/antibiotic combinations, showing that this is independent of the original architecture of the biofilm and the manner in which the critical decrease is achieved. There are three instances where the reduction at 4 h is beyond the critical reduction but not significantly different than the sum of component reductions (ST biofilms shocked at 60 °C for 5 min or 70 °C for 1 min and DFR biofilms shocked at 70 °C for 2 min, all exposed to 4 μg/mL ciprofloxacin), but in all three cases, the biofilms proceed to die off within 24 h, resulting in a population reduction that is again much larger than the sum of the heat-shock only and antibiotics-only reductions. While neither treatment by itself achieved the critical population drop, the combination of the treatments did, prompting further population decrease, which makes the treatment appear synergistic, even though there is no indication that the treatments actually interact in any way.

In summary, previous studies have suggested a synergistic link between antibiotic exposure and thermal shock in the eradication of bacterial biofilms, i.e., that one treatment enhances the efficacy of the other in some way, resulting in a population reduction larger than predicted from simply adding the effects of the treatments by themselves. This study investigated that link, growing *P aeruginosa* biofilms of two disparate population densities and architectures and combining a variety of different thermal shocks with different concentrations of ciprofloxacin. When the sum of the log population decrease for thermal shock alone and for antibiotics alone was less than four orders of magnitude, the population decrease when combining the two treatments was not significantly different than the sum from the individual treatments, indicating no synergism, just orthogonal mechanisms. However, when the sum from the individual treatments was more than four orders of magnitude, the decrease from the combined treatment was even larger, eventually eliminating the biofilm altogether. While this gives the appearance of synergistic interaction in just those instances, the same eventual elimination is observed even without antibiotic exposure when the initial population decrease exceeds four orders of magnitude from thermal shock alone. It appears likely that thermal shock and antibiotics act with strictly orthogonal mechanisms on *P aeruginosa* biofilms, but both contribute to a separate, general, critical decrease phenomenon.

4. Materials and Methods

4.1. Streak and Inoculum

Pseudomonas aeruginosa PAO1 (15692, American Type Culture Collection, Manassas, VA, USA) was streaked on an agar plate (Difco Nutrient Agar, Sparks, MD, USA) and incubated inverted for 24 h at 37 °C. Two colonies from the streaked plate were harvested and moved into 5 mL (30 g/L) Tryptic Soy Broth (TSB, Becton, Dickinson and Company, Franklin Lakes, NJ, USA) using a sterile inoculating loop. Inoculated TSB was incubated for 24 h at 37 °C to form an inoculum with an average of ~10^9 colony forming units (CFU) per mL.

4.2. Biofilm Culture

4.2.1. Shaker Table Biofilm

In total, 0.333 mL of inoculum and 5 mL of 30 g/L TSB were added to each well of 4-well dishes (Thermo Fisher Scientific, Waltham, MA, USA). In each well, a microscope slide fully frosted on one side, 75 mm × 25 mm × 1 mm (Leica Biosystems, Buffalo Grove, IL, USA) was immersed, and then the dish was sealed with parafilm. Dishes were placed on an orbital shaker table (VWR 1000, 15 mm orbit, Radnor, PA, USA) set at 160 rpm, and placed inside an incubator at 37 °C for 96 h to culture mature ST biofilms.

4.2.2. Drip Flow Reactor (DFR) Biofilm

DFR biofilm culture includes two distinct modes, batch and continuous. A frosted microscope slide was immersed in 15 mL (30 g/L) TSB in each well of a 4-well reactor (Biosurface Technologies Corporation, Bozeman, MT). In total, 1 mL of inoculum was added to each well, then the wells were sealed with their lids. During batch mode, the reactor was at rest horizontally inside a 37 °C incubator for 4 h. To begin continuous mode, the reactor was tilted by 10 degrees and a steady drip of TSB (3 g/L) was applied to the upper end of the nascent biofilm, flowing down the slide by gravity to a drain hose at the lower end. The continuous mode ran for 20 h inside the incubator at 37 °C, with a drip flowrate of 1.25 L/day per well.

4.3. Thermal Shock

Mature ST or DFR biofilms were removed from their culture wells and rinsed in 5 mL of 3 g/L TSB for 1 min to remove planktonic bacteria, and then transferred to preheated 4-well dishes with 5 mL of 3 g/L TSB in each well. Biofilms were thermally shocked at (50, 60, or 70 °C) for (1, 2, 5, 10, or 30 min). Temperature was controlled by keeping the dishes in a water bath at the target temperature for 30 min prior to thermal shock and throughout the shock itself. Biofilms were transferred immediately after the thermal shock to new 4-well dishes with 5 mL of 3 g/L TSB per well at ambient temperature.

4.4. Re-Incubation

To investigate the viability of thermally shocked biofilms, they were re-incubated under conditions identical to their initial culturing. Thermally shocked DFR biofilms were placed in a fresh DFR inclined at 10°, in which the 3 g/L TSB drip. Biofilms were re-incubated for 4 or 24 h at 37 °C and the same flowrate of 1.25 L/day per well.

4.5. Antibiotic Exposure

In total, 5 mg/mL ciprofloxacin (CP) stock was prepared by dissolving ciprofloxacin hydrochloride (MP Biomedicals, Santa Ana, CA, USA) in de-ionized water. The stock was filtered with a 0.22 μm PES membrane sterile filter (Millex®GP filter unit) and stored at 2 °C.

4.5.1. Shaker Table Antibiotic Exposure

Mature ST biofilms were rinsed in 5 mL of TSB (3 g/L) for 1 min to remove planktonic bacteria, and then placed in a fresh 4-well dish, where each well contained 5 mL TSB (30 g/L) and CP (0.25, 1, 4, 16, 64, or 256 μg/mL). These concentrations were selected to observe antibiotic effects below, at, or above intravenous administration concentrations (4 μg/mL) on biofilm [39]. Biofilms were kept in these antibiotics challenge plates for 4, 24, or 48 h at 37 °C.

4.5.2. Drip Flow Reactor Antibiotic Exposure

Mature DFR biofilms were placed in a fresh DFR inclined at 10°, in which the 3 g/L TSB drip also contained CP at 0.25, 4, or 64 μg/mL. The flowrate remained at 1.25 L/day per well for 4, 24, or 48 h at 37 °C.

4.6. Antibiotic and Thermal Exposure

To investigate the interaction of antibiotics and thermal shock on the reduction of biofilm population density, mature biofilms were exposed to antibiotics for four hours as described above. Following this exposure, biofilms were immediately transferred to preheated 4-well dishes with 5 mL of 3 g/L TSB and the same antibiotic concentration for thermal shock as described above. After the thermal shock, biofilms were either immediately enumerated, or reincubated in a new 4-well dish (ST) or new reactor (DFR) at 37 °C with the same antibiotics concentration for the remainder of a 24 or 48 h antibiotics exposure.

4.7. Enumeration

Biofilms' population density was determined via suspension and plating. Biofilms were transferred to a fresh 4-well dish containing 5 mL of 3 g/L TSB and mechanically disrupted and homogenized with a sonicator bath for 10 min at 45kHz (VWR Symphony, 9.5 L). The sonicated homogenous solutions were then serial diluted by 10 fold, and spot plated with 10 µL samples on nutrient agar plates. After 20 h of incubation, grown colonies were counted and the population density of colony forming units per square centimeter was calculated.

4.8. Confocal Microscopy

Confocal laser scanning microscopy was performed to evaluate ST and DFR biofilms' architectural characteristics. Biofilms were exposed to fluorescent dyes from a LIVE/DEAD BacLight Bacterial Viability Kit (Invitrogen, Eugene, OR) for 20 min in low-light conditions to stain the biofilms prior to imaging with a non-inverted confocal microscope (Zeiss LSM 710, Oberkochen, Germany).

4.9. Statistical Analysis

Enumeration results were calculated on a log scale for statistical analysis. Significance was determined using two-tailed student-t tests with a 95% confidence interval. Variance for each trial arm is assumed to be uncorrelated, and differences in variance were reconciled per Cochran [41].

5. Conclusions

Reincubation studies on treated *P aeruginosa* biofilms indicate a critical population decrease beyond which the population will continue decreasing rather than recover. The factor by which thermal shock or ciprofloxacin reduce biofilm population density appear to be the same regardless of whether the other treatment factor is also applied, indicating that the reduction mechanisms are orthogonal (i.e., independent of each other) rather than synergistic. However, beyond the critical overall population decrease factor, the biofilm population continues to crash to zero, resulting in a post-treatment reduction far beyond the product of the individual treatments, making them appear synergistic even if their mechanisms do not actually overlap.

Author Contributions: Conceptualization, H.A. and E.N.; Data curation, H.A.; Formal analysis, H.A. and E.N.; Funding acquisition, H.A. and E.N.; Investigation, H.A., Y.G. and H.C.; Methodology, H.A. and E.N.; Project administration, E.N.; Resources, H.A. and E.N.; Supervision, E.N.; Validation, H.A.; Visualization, H.A.; Writing—original draft, H.A. and E.N.; Writing—review and editing, E.N. All authors have read and agreed to the published version of the manuscript.

Funding: This research was funded in part by the American Heart Association, grant number 18IPA34170108 and the National Science Foundation, grant number CBET-1133297.

Data Availability Statement: The raw data collated and reported here is available in the NuxollResearch-Group DataVerse (URL and journal-specific dataset name to be added after manuscript acceptance).

Acknowledgments: Aljaafari, Haydar is a PhD student supported by the Higher Committee of Education Development in Iraq (HCED). The Zeiss LSM 710 confocal microscope was also obtained with support from the National Institutes of Health (1 S10 RR025439–01).

Conflicts of Interest: The authors declare no conflict of interest, and the funders had no role in the design of the study; in the collection, analyses, or interpretation of data; in the writing of the manuscript, or in the decision to publish the results.

References

1. McDermott, K.W.; Freeman, W.J.; Elixhauser, A. *Overview of Operating Room Procedures during Inpatient Stays in U.S. Hospitals, 2014*; HCUP Statistical brief #233; Agency for Healthcare Research and Quality: Rockville, MD, USA, 2017.
2. O'Toole, P.; Maltenfort, M.G.; Chen, A.F.; Parvizi, J. Projected increase in periprosthetic joint infections secondary to rise in diabetes and obesity. *J. Arthroplast.* **2016**, *31*, 7–10. [CrossRef]
3. Parvizi, J.; Ghanem, E.; Azzam, K.; Davis, E.; Jaberi, F.; Hozack, W. Periprosthetic infection: Are current treatment strategies adequate? *Acta Orthop. Belg.* **2008**, *74*, 793–800.
4. Phillips, J.E.; Crane, T.P.; Noy, M.; Elliott, T.S.J.; Grimer, R.J. The incidence of deep prosthetic infections in a specialist orthopaedic hospital. *J. Bone Jt. Surg. Br. Vol.* **2006**, *88*, 943–948. [CrossRef]
5. Cui, Q.; Mihalko, W.M.; Shields, J.S.; Ries, M.; Saleh, K.J. Antibiotic-impregnated cement spacers for the treatment of infection associated with total hip or knee arthroplasty. *J. Bone Jt. Surg. Am. Vol.* **2007**, *89*, 871–882. [CrossRef]
6. Anderl, J.N.; Zahller, J.; Roe, F.; Stewart, P. Role of nutrient limitation and stationary-phase existence in klebsiella pneumoniae biofilm resistance to ampicillin and ciprofloxacin. *Antimicrob. Agents Chemother.* **2003**, *47*, 1251–1256. [CrossRef]
7. Piddock, L.J.V. Multidrug-resistance efflux pumps? Not just for resistance. *Nat. Rev. Genet.* **2006**, *4*, 629–636. [CrossRef] [PubMed]
8. Liu, Y.; Kamesh, A.C.; Xiao, Y.; Sun, V.; Hayes, M.; Daniell, H.; Koo, H. Topical delivery of low-cost protein drug candidates made in chloroplasts for biofilm disruption and uptake by oral epithelial cells. *Biomaterials* **2016**, *105*, 156–166. [CrossRef] [PubMed]
9. Hall, C.W.; Mah, T.-F. Molecular mechanisms of biofilm-based antibiotic resistance and tolerance in pathogenic bacteria. *FEMS Microbiol. Rev.* **2017**, *41*, 276–301. [CrossRef]
10. Pérez-Díaz, M.; Alvarado-Gomez, E.; Magaña-Aquino, M.; Sánchez-Sánchez, R.; Velasquillo, C.; Gonzalez, C.; Ganem-Rondero, A.; Martinez-Castanon, G.-A.; Zavala-Alonso, N.; Martinez-Gutierrez, F. Anti-biofilm activity of chitosan gels formulated with silver nanoparticles and their cytotoxic effect on human fibroblasts. *Mater. Sci. Eng. C* **2016**, *60*, 317–323. [CrossRef]
11. Darouiche, R.O. Treatment of infections associated with surgical implants. *N. Engl. J. Med.* **2004**, *350*, 1422–1429. [CrossRef]
12. Tran, N.; Tran, P. Nanomaterial-based treatments for medical device-associated infections. *ChemPhysChem* **2012**, *13*, 2481–2494. [CrossRef] [PubMed]
13. Osmon, D.R.; Berbari, E.F.; Berendt, A.R.; Lew, D.; Zimmerli, W.; Steckelberg, J.M.; Rao, N.; Hanssen, A.; Wilson, W.R. Executive summary: Diagnosis and management of prosthetic joint infection: Clinical practice guidelines by the Infectious Diseases Society of America. *Clin. Infect. Dis.* **2013**, *56*, 1–10. [CrossRef] [PubMed]
14. Berend, K.R.; Lombardi, A.V.; Morris, M.J.; Bergeson, A.G.; Adams, J.; Sneller, M.A. Two-stage treatment of hip periprosthetic joint infection is associated with a high rate of infection control but high mortality. *Clin. Orthop. Relat. Res.* **2013**, *471*, 510–518. [CrossRef] [PubMed]
15. Cooper, H.; Della Valle, C.J. The two-stage standard in revision total hip replacement. *Bone Jt. J.* **2013**, *95-B*, 84–87. [CrossRef]
16. Rohde, H.; Burandt, E.C.; Siemssen, N.; Frommelt, L.; Burdelski, C.; Wurster, S.; Scherpe, S.; Davies, A.; Harris, L.; Horstkotte, M.A.; et al. Polysaccharide intercellular adhesin or protein factors in biofilm accumulation of Staphylococcus epidermidis and Staphylococcus aureus isolated from prosthetic hip and knee joint infections. *Biomaterials* **2007**, *28*, 1711–1720. [CrossRef]
17. Wilkins, M.; Hall-Stoodley, L.; Allan, R.; Faust, S.N. New approaches to the treatment of biofilm-related infections. *J. Infect.* **2014**, *69*, S47–S52. [CrossRef]
18. Tulloch, A.W.; Chun, Y.; Kealey, C.; Mohanchandra, K.P.; Chang, J.; Milisavljevic, V.; Levi, D.S.; Lawrence, P.F.; Rigberg, D.A. PS236. Hydrophilic surface treatment of thin film nickel titanium reduces bacterial biofilm production compared to commercially available endograft materials. *J. Vasc. Surg.* **2010**, *51*, 79S–80S. [CrossRef]
19. Wei, J.; Ravn, D.B.; Gram, L.; Kingshott, P. Stainless steel modified with poly(ethylene glycol) can prevent protein adsorption but not bacterial adhesion. *Colloids Surf. B Biointerfaces* **2003**, *32*, 275–291. [CrossRef]
20. Tunney, M.; Gorman, S. Evaluation of a poly(vinyl pyrollidone)-coated biomaterial for urological use. *Biomaterials* **2002**, *23*, 4601–4608. [CrossRef]
21. Francolini, I.; Donelli, G. Prevention and control of biofilm-based medical-device-related infections. *FEMS Immunol. Med. Microbiol.* **2010**, *59*, 227–238. [CrossRef]
22. Koopaie, M.; Bordbar-Khiabani, A.; Kolahdooz, S.; Darbandsari, A.K.; Mozafari, M. Advanced surface treatment techniques counteract biofilm-associated infections on dental implants. *Mater. Res. Express* **2020**, *7*, 015417. [CrossRef]
23. Chen, J.; Howell, C.; Haller, C.A.; Patel, M.S.; Ayala, P.; Moravec, K.A.; Dai, E.; Liu, L.; Sotiri, I.; Aizenberg, M.; et al. An immobilized liquid interface prevents device associated bacterial infection in vivo. *Biomaterials* **2017**, *113*, 80–92. [CrossRef]
24. Khare, M.D.; Bukhari, S.S.; Swann, A.; Spiers, P.; McLaren, I.; Myers, J. Reduction of catheter-related colonisation by the use of a silver zeolite-impregnated central vascular catheter in adult critical care. *J. Infect.* **2007**, *54*, 146–150. [CrossRef]
25. Tan, L.; Li, J.; Liu, X.; Cui, Z.; Yang, X.; Zhu, S.; Li, Z.; Yuan, X.; Zheng, Y.; Yeung, K.; et al. Rapid biofilm eradication on bone implants using red phosphorus and near-infrared light. *Adv. Mater.* **2018**, *30*, e1801808. [CrossRef] [PubMed]
26. Wang, H.; Ren, D. Controlling Streptococcus mutans and Staphylococcus aureus biofilms with direct current and chlorhexidine. *AMB Express* **2017**, *7*, 1–9. [CrossRef] [PubMed]

27. Coffel, J.; Nuxoll, E. Magnetic nanoparticle/polymer composites for medical implant infection control. *J. Mater. Chem. B* **2015**, *3*, 7538–7545. [CrossRef] [PubMed]
28. Van der Borden, A.; van der Mei, H.; Busscher, H. Electric block current induced detachment from surgical stainless steel and decreased viability of Staphylococcus epidermidis. *Biomaterials* **2005**, *26*, 6731–6735. [CrossRef]
29. Carmen, J.C.; Roeder, B.L.; Nelson, J.L.; Ogilvie, R.L.R.; Robison, R.; Schaalje, G.B.; Pitt, W.G. Treatment of biofilm infections on implants with low-frequency ultrasound and antibiotics. *Am. J. Infect. Control.* **2005**, *33*, 78–82. [CrossRef] [PubMed]
30. Zhang, G.; Zhang, X.; Yang, Y.; Zhang, H.; Shi, J.; Yao, X. Near-infrared light-triggered therapy to combat bacterial biofilm infections by MoSe2 /TiO2 nanorod arrays on bone implants. *Adv. Mater. Interfaces* **2019**, *7*. [CrossRef]
31. Richardson, I.P.; Sturtevant, R.; Heung, M.; Solomon, M.; Younger, J.G.; Vanepps, J.S. Hemodialysis catheter heat transfer for biofilm prevention and treatment. *ASAIO J.* **2016**, *62*, 92–99. [CrossRef]
32. Bandara, H.; Nguyen, D.; Mogarala, S.; Osiñski, M.; Smyth, H. Magnetic fields suppress Pseudomonas aeruginosa biofilms and enhance ciprofloxacin activity. *Biofouling* **2015**, *31*, 443–457. [CrossRef]
33. O'Toole, A.; Ricker, E.B.; Nuxoll, E. Thermal mitigation of Pseudomonas aeruginosabiofilms. *Biofouling* **2015**, *31*, 665–675. [CrossRef]
34. Chopra, R.; Shaikh, S.; Chatzinoff, Y.; Munaweera, I.; Cheng, B.; Daly, S.M.; Xi, Y.; Bing, C.; Burns, D.; Greenberg, D.E. Employing high-frequency alternating magnetic fields for the non-invasive treatment of prosthetic joint infections. *Sci. Rep.* **2017**, *7*, 1–14. [CrossRef] [PubMed]
35. Ricker, E.B.; Nuxoll, E. Synergistic effects of heat and antibiotics on Pseudomonas aeruginosa biofilms. *Biofouling* **2017**, *33*, 855–866. [CrossRef] [PubMed]
36. Hajdu, S.; Holinka, J.; Reichmann, S.; Hirschl, A.M.; Graninger, W.; Presterl, E. Increased temperature enhances the antimicrobial effects of daptomycin, vancomycin, tigecycline, fosfomycin, and cefamandole on staphylococcal biofilms. *Antimicrob. Agents Chemother.* **2010**, *54*, 4078–4084. [CrossRef] [PubMed]
37. Wu, T.; Wang, L.; Gong, M.; Lin, Y.; Xu, Y.; Ye, L.; Yu, X.; Liu, J.; Liu, J.; He, S.; et al. Synergistic effects of nanoparticle heating and amoxicillin on H. pylori inhibition. *J. Magn. Magn. Mater.* **2019**, *485*, 95–104. [CrossRef]
38. Pijls, B.G.; Sanders, I.M.J.G.; Kuijper, E.J.; Nelissen, R.G.H.H. Synergy between induction heating, antibiotics, and N-acetylcysteine eradicates Staphylococcus aureus from biofilm. *Int. J. Hyperth.* **2020**, *37*, 130–136. [CrossRef]
39. McEvoy, G.K. (Ed.) *AHFS Drug Information 2008*; American Society of Health-System Pharmacists: Bethesda, MD, USA, 2008.
40. Ricker, E.B.; Aljaafari, H.A.S.; Bader, T.M.; Hundley, B.S.; Nuxoll, E. Thermal shock susceptibility and regrowth of Pseudomonas aeruginosabiofilms. *Int. J. Hyperth.* **2018**, *34*, 168–176. [CrossRef] [PubMed]
41. Cochran, W.J. Approximate significance levels of the behrens-fischer test. *Biometrics* **1964**, *20*, 191–195. [CrossRef]

Article

New Functionalized Macroparticles for Environmentally Sustainable Biofilm Control in Water Systems

Ana C. Barros [1,*], Ana Pereira [1], Luis F. Melo [1] and Juliana P. S. Sousa [2]

1 LEPABE—Laboratory for Process Engineering, Environment, Biotechnology and Energy, Faculty of Engineering, University of Porto, 4200-465 Porto, Portugal; aalex@fe.up.pt (A.P.); lmelo@fe.up.pt (L.F.M.)
2 INL, International Iberian Nanotechnology Laboratory, Avenida Mestre José Veiga s/n, 4715-330 Braga, Portugal; juliana.sousa@inl.int
* Correspondence: acbarros@fe.up.pt; Tel.: +351-22-508-3603

Abstract: Reverse osmosis (RO) depends on biocidal agents to control the operating costs associated to biofouling, although this implies the discharge of undesired chemicals into the aquatic environment. Therefore, a system providing pre-treated water free of biocides arises as an interesting solution to minimize the discharge of chemicals while enhancing RO filtration performance by inactivating bacteria that could form biofilms on the membrane system. This work proposes a pretreatment approach based on the immobilization of an industrially used antimicrobial agent (benzalkonium chloride—BAC) into millimetric aluminum oxide particles with prior surface activation with DA—dopamine. The antimicrobial efficacy of the functionalized particles was assessed against *Escherichia coli* planktonic cells through culturability and cell membrane integrity analysis. The results showed total inactivation of bacterial cells within five min for the highest particle concentration and 100% of cell membrane damage after 15 min for all concentrations. When reusing the same particles, a higher contact time was needed to reach the total inactivation, possibly due to partial blocking of immobilized biocide by dead bacteria adhering to the particles and to the residual leaching of biocide. The overall results support the use of Al_2O_3-DA-BAC particles as antimicrobial agents for sustainable biocidal applications in continuous water treatment systems.

Keywords: biocidal particles; functionalization; benzalkonium chloride; *Escherichia coli*; antimicrobial activity

Citation: Barros, A.C.; Pereira, A.; Melo, L.F.; Sousa, J.P.S. New Functionalized Macroparticles for Environmentally Sustainable Biofilm Control in Water Systems. *Antibiotics* **2021**, *10*, 399. https://doi.org/10.3390/antibiotics10040399

Academic Editor: Carlos M. Franco

Received: 2 March 2021
Accepted: 3 April 2021
Published: 7 April 2021

1. Introduction

The so-called clean water crisis, which already affects near four billion people [1], is a major humanitarian issue. The effect of unbalanced urbanization, population growth and inadequate use of water is expected to increase between 22% and 34% until 2050 [1]. The requalification or reuse of water from non-potable sources is considered one of the most promising and attractive ways to face such water crisis, with special emphasis on the use of Reverse Osmosis (RO) systems to accomplish proper water "purification".

Biofouling—the attachment and accumulation of unwanted microorganisms in the form of a biofilm to an extent that interferes negatively with the filtration process [2]—is a major factor contributing to the decrease of the operational performance of RO systems. It is widely consensual that biofouling in RO systems is the main responsible for the decrease in permeate flow rate and quality, increase of pressure drop and of energy demand [3]. Since biofilm build-up is still practically unavoidable, new and sustainable approaches are needed to circumvent this issue [4].

The use of antimicrobial agents, typically disinfectants and/or biocides, is a challenging question across the water treatment sector. On one hand, antimicrobial agents per se are not always able to eradicate biofouling, nor have a similar killing effect on planktonic and sessile (attached) bacteria [5,6]. Additionally, the use of antimicrobial agents dissolved in the water is known to have a negative impact on process economy, health and environment, mostly associated with their discharge along with the rejection stream [7]. On the other hand, it is generally accepted that biocides are important to control microbiological proliferation. They became a "must have" feature to include in RO systems and are used not only in corrective procedures but, also, in routine-based operations [8].

This paradigm opened an opportunity to explore different ways of keeping systems microbiologically controlled while reducing the environmental and public health impacts associated with the use of biocides as a daily-basis procedure. The rise of nanotechnology and the possibility to aggregate new physicochemical properties to the materials in terms of surface area, topography and chemical function [9] is being extensively explored to fulfil this challenge. Several of the approaches that are being investigated include the development of materials with self-antimicrobial properties—e.g., nano-Ag, nano-ZnO, etc. [10–18]—and the immobilization of biocidal agents on nanostructured materials due to their high surface-area-to-volume ratio and surface modifiability [19–22]. These materials can be used as adsorbents of pollutants or as carriers for the biocidal agent. However, serious environmental and health concerns are arising about the fate and toxicity of nanoparticles in water systems [23].

In the present work, a different approach is proposed: (i) the antimicrobial particles are pellets with overall millimetric size, which can be packed in a particle bed system through which the water containing microorganisms flows continuously; (ii) due to their relatively larger dimensions (10^6 times the size of nanomaterials), these particles can be easily contained in a vessel and not carried away with the outlet water stream. As regards the composition of the particles, metal oxides were chosen, since they have already been successfully applied for environmental monitoring, remediation and pollution prevention [24] and have a low cost. Commercial metal oxides, namely γ-alumina, present excellent mechanical and chemical properties; however, their surface chemical properties do not favor the direct immobilization of biocides. Therefore, it is necessary to previously activate the surfaces to increase their surface reactivity. Herein, an easy and reproducible approach to immobilize biocides onto the surface of metal oxides is proposed. First, the metal oxide particles are functionalized with a precursor to activate their surface reactivity and then the biocide is immobilized onto the surface by putting the metal oxide particles in contact with the biocidal solution.

In this work, alumina particles were functionalized by using a nitrogen precursor (Dopamine (DA)). Subsequently, the biocide benzalkonium chloride (BAC) was immobilized in these particles, and their antimicrobial activity was assessed against *Escherichia coli* bacteria in planktonic state. BAC is a cationic surfactant, belonging to the family of QACs (quaternary ammonium compounds), with antibacterial properties often used in water treatment [25]. BAC gathered the attention of the scientific community because of its physical–chemical characteristics [26] and its mechanisms of interaction with bacteria [27], later discussed in the Results and Discussion section of the present paper. Several studies regarding the immobilization of QAC compounds in nanostructured material were published [22], including the functionalization of metal oxide nanoparticles [28,29]. However, none of those cases include the functionalization of macroscopic metal oxide pellets.

The present work is the first step to pursue the incorporation of antimicrobial macroscopic particles into a continuous flow bed-reactor (fixed or fluidized) to be added as an RO pretreatment step of a water stream. The success of this approach will contribute to optimize the application of biocides while minimizing the environmental and health impacts. Therefore, the solution here proposed considers the use of commercially available metal oxides (pellet shape, overall average dimension of 3 mm) with added biocidal properties, overcoming two issues brought by real-field applications of nanoparticles: the difficulty in removing nanoparticles from industrial-scale water streams and the high pressure drop caused by the compaction of a bed of nanoparticles.

2. Results and Discussion

2.1. Functionalization Characteristics

In order to activate the surface of alumina samples, the commercial materials were functionalized with a nitrogen precursor, namely dopamine. The nitrogen groups introduced in the particles' surface act as anchorage points to immobilize the biocide. The prepared materials were characterized using different techniques, yielding the results described in the next subtopics.

2.1.1. Thermal Stability of Functionalized Metal Oxides

To get more insight into the packing density, thermogravimetric analysis was used to estimate the amount of functional groups anchored to the surface of the particles (Figure 1). By observing the weight loss profiles of the raw material (Al_2O_3), it appears that the latter does not have organic impurities, because the values of burn-off obtained for these samples are very low (3%), see Figure 1a.

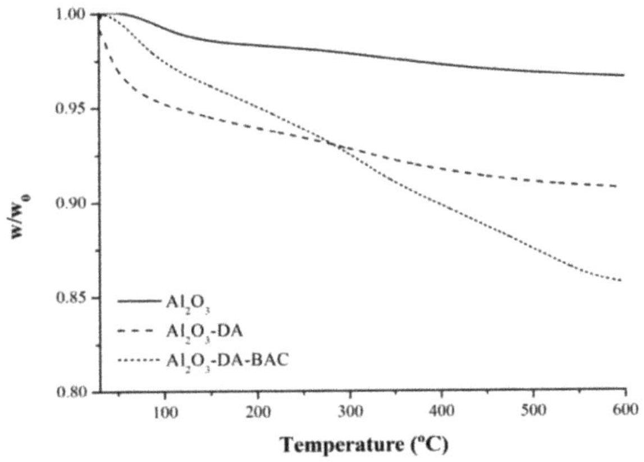

(**a**)

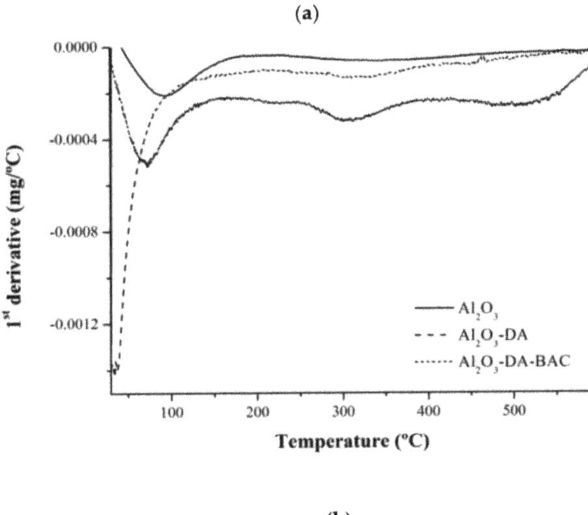

(**b**)

Figure 1. Thermogravimetric analysis, TG (**a**) and first derivative curve, DTG (**b**) of functionalized and Al_2O_3 particles.

The functionalization with nitrogen precursors increased the values of the burn-off. The sample Al_2O_3-DA present weight losses between 200 and 400 °C of around 2%. The thermogravimetric analysis (Figure 1a) confirmed the success of the functionalization with nitrogen precursors. The interaction between dopamine and inorganic materials is most likely due to hydrogen bonding, conferring a high thermal stability to the sample.

The nitrogen groups introduced in alumina particles by dopamine will be the anchor points for BAC immobilization. It can be seen in Figure 1b that the Al_2O_3-DA-BAC sample presents a total mass loss of 3% in the temperature ranges of 280–390 °C and 426–598 °C. This 3% corresponds to the amount of immobilized BAC per unit mass of particles, as described in Section 3.4.3.

2.1.2. Adsorption/Desorption Isotherms

Adsorption and desorption isotherms of the particles were performed. As can be seen in Figure 2, all particles showed a type IV isotherm behavior, which indicates the presence of mesopores [30–32].

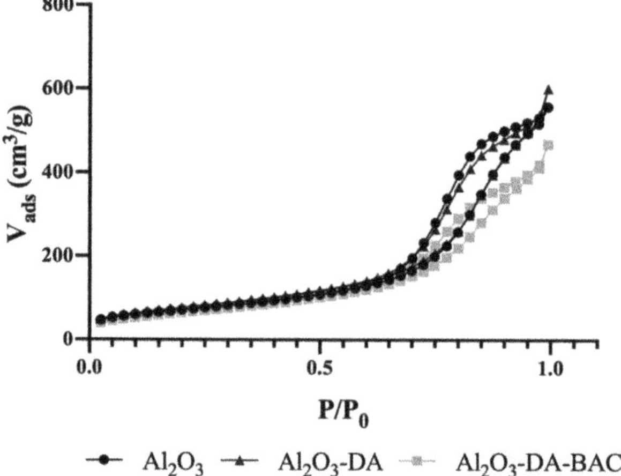

Figure 2. Nitrogen adsorption and desorption isotherms of functionalized and Al_2O_3 particles.

Table 1 summarizes the results obtained from nitrogen adsorption–desorption isotherms (Figure 2). According to the N_2 physisorption, it appears that the functionalization treatment changed the textural properties of alumina materials. The surface area of the sample treated with dopamine increases after the treatment (from 258 to 280 m^2/g). The development of porosity during the treatment with dopamine (Table 1), observed for the alumina materials, is due to the widening of existing pores (d_p increased from 8.2 to 9.0 nm), confirming the successful functionalization of this material. This can be explained by the fact that the dopamine solution has a high pH value (basic solution), which can cause some dissolution/etching of the aluminum oxide. After the BAC immobilization, the surface area of this sample decreased to 240 m^2/g which can be explained by the occupation of some pores by the BAC molecules, since the pore volume decreased from 0.77 to 0.60 cm^3/g.

Table 1. Surface area parameters obtained for the functionalized particles and the control (Al_2O_3).

Sample	S_{BET} (m^2/g)	d_p (nm)	$V_{p0:0.95}$ (cm^3/g)
Al_2O_3	258	8.2	0.77
Al_2O_3-DA	280	9.0	0.77
Al_2O_3-DA-BAC	240	8.1	0.60

S_{BET}—surface area, d_p—pore diameter and $V_{p0:0.95}$—pore volume.

2.1.3. X-ray Diffraction (XRD) Analysis

The XRD pattern of the alumina particles is represented in Figure 3. The original alumina patterns show relatively strong peaks at 2θ values at 28, 38, 49, 65 and 72°, which are attributed to the reflections of Boehmite (γ-AlO(OH)). The DA-functionalized sample presents a crystalline phase of η- Al_2O_3, detected by the presence of peaks at 37, 39, 46 and 67°. Additionally, Al_2O_3-DA-BAC particles show an entire transformation to γ-Al_2O_3, detected by the presence of two peaks at 47 and 68°. Data from the XRD also confirms the successful functionalization of the particles with biocide.

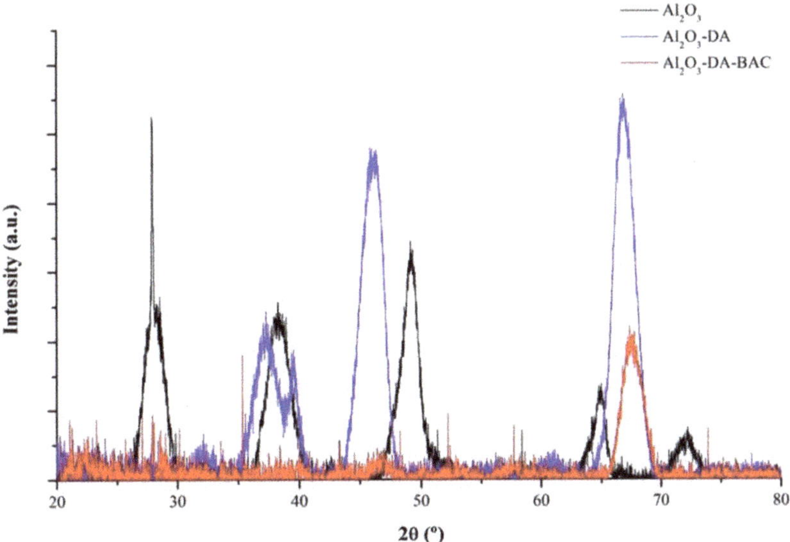

Figure 3. X-ray diffraction (XRD) pattern of functionalized and Al_2O_3 particles.

2.1.4. Point of Zero Charge—pH_{PZC}

Figure 4 shows the pH_{PZC} values estimated for the alumina samples (prior to functionalization), which is found to be 5.7. This value increases to 7.3 after introduction of –NH_2 groups, suggesting that the metal oxide surfaces have been modified with positively charged amino groups.

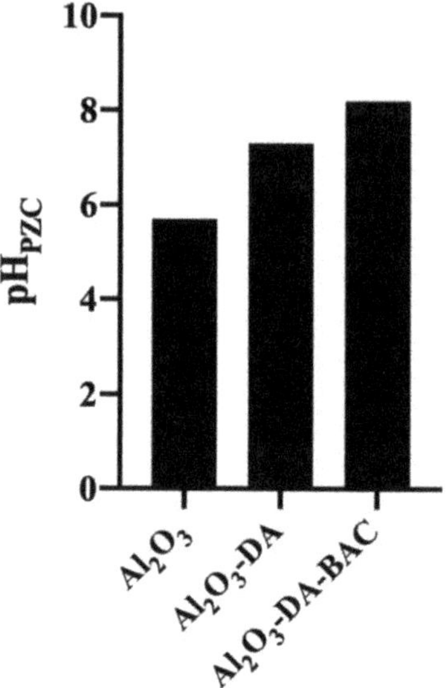

Figure 4. Point of zero charge of the functionalized and Al_2O_3 particles.

Furthermore, after BAC incorporation, the pH_{PZC} raised to 8.2. The basic nature of benzalkonium chloride corroborates the pH_{PZC} increase.

2.1.5. Fourier-Transform Infrared Spectroscopy (FTIR) Analysis

FTIR spectra of Al_2O_3 particles (Figure 5) show a band in the region of 3300–3700 cm^{-1}, which corresponds to the water O-H stretch [33]. Ribena [34] also found bands in the region 3000–3400 cm^{-1} after coating the surface of its particles with dopamine and attributed it to the intermolecular hydrogen bonds (O-H, N-H and aromatic CH2 stretching vibrations) naturally occurring between dopamine molecules. Regarding Figure 5, another band was found in the region of 1580–1650 cm^{-1} for Al_2O_3-DA and Al_2O_3-DA-BAC particles. This band corresponds to N-H bending [33], meaning that functionalization with dopamine coated the particles with amine groups. The functionalization of particles with BAC resulted in the appearance of two small peaks at 1153 and 1211 cm^{-1} on the curve in red. These peaks are associated to C-N and N-CH_3 stretching vibrations, which are characteristic of quaternary amines [35,36].

FTIR results proved again the successful functionalization of particles with nitrogen and biocidal groups.

The overall physical and chemical characterizations discussed above (BAC peaks in the FTIR diagram, mass loss in the TG curves, XRD analysis and the sorption isotherms) point-out to a successful immobilization of active BAC groups after the proposed functionalization.

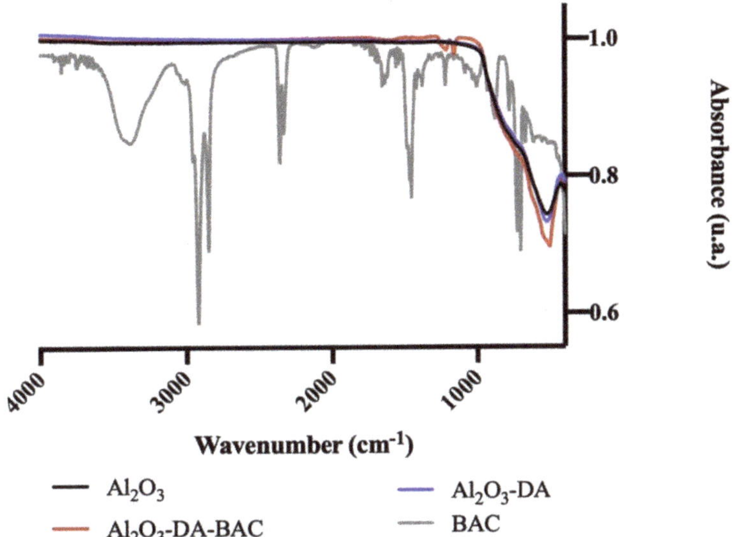

Figure 5. Fourier-Transform Infrared Spectroscopy (FTIR) spectra of functionalized and Al_2O_3 particles. Benzalkonium chloride (BAC) in solution was used as control.

2.2. Particles' Antimicrobial Activity

To assess the *Escherichia coli* susceptibility to the immobilized biocide, zones of inhibition were determined. Data shown in Table 2 demonstrate that Al_2O_3 and Al_2O_3-DA particles do not induce any zone of inhibition, suggesting that the core materials without biocide, for the present test conditions, have a very small or no antibacterial effect on *E. coli*. On the other hand, Al_2O_3-DA-BAC particles exhibited a zone of inhibition diameter of 18 ± 2 mm, which implies that there is some biocide leaching from the particles. Hassan and Elbagoury [37], while studying the susceptibility of some bacterial isolates to BAC in solution, observed similar inhibition diameters. These authors found zone of inhibition average diameters of 13 and 17 mm when *Pseudomonas* spp. cells were exposed to a 2% BAC solution.

Table 2. Zone of inhibition diameter obtained for the tested particles against *Escherichia coli*.

	Zone of Inhibition (mm)
Al_2O_3	0 ± 0
Al_2O_3-DA	0 ± 0
Al_2O_3-DA-BAC	18 ± 2

Values are presented as the mean ± standard deviation of two independent experiments.

The formation of a zone of inhibition is related to the diffusion constant of the antimicrobial agent [38–40] and, therefore, can be related to its amount and motility. According to the same authors [38–40], a zone of inhibition appears when the concentration of an antimicrobial agent exceeds the minimum inhibitory concentration or its critical concentration [38–41]. Although, this test highly depends on several experimental factors (e.g., test medium, organism and its microbial concentration and biocide type), it is expected that for the same experimental conditions, zones of inhibition can provide a qualitative indication of the bacterial susceptibility against a given antimicrobial agent [42]. Therefore, the results in Table 2 suggest that the functionalized particles (Al_2O_3-DA-BAC) have a strong antibacterial activity potential.

2.3. *Effect of Immobilized Biocide Concentration on the Particles' Antimicrobial Activity*

According to what was described in the Materials and Methods section, different ratios of particles per unit volume of solution, summarized in Table 3, were tested. It is important to note that the concentration of immobilized biocide per unit mass of each particle was 3% (*w*/*w*) of BAC, according to the data obtained in the thermogravimetric analysis (as described in Section 2.1.1). Therefore, the tested concentrations of 500, 1000 and 3000 mg/L refer to the "overall concentration of immobilized biocide" defined as mg of immobilized biocide per L of total liquid volume (bacterial suspension). Thus, the different concentrations correspond to different amounts of particles per unit liquid volume.

Table 3. Tested conditions regarding the overall immobilized biocide concentration.

Particles	Immobilized BAC *w*/*w* (%): Mass of BAC (g) per 100 g of Particles	Ratio Mass of Particles (g)/Volume of Solution (L)	Overall Concentration under Test (mg of Immobilized Biocide/L Solution)
Al_2O_3-DA-BAC	3	16.8	500
		33.5	1000
		100.3	3000

The antimicrobial effect of such functionalized particles (Al_2O_3-DA-BAC), considering the different overall concentrations of immobilized biocide, was evaluated according to the bacteria culturability and membrane integrity (propidium iodide, PI, uptake percentage) criteria.

2.3.1. Culturability

The reduction of the colony forming units per milliliter (CFU/mL) of *E. coli* planktonic cells over time, exposed to different overall concentrations of Al_2O_3-DA-BAC particles, is shown in Figure 6a. In this figure, it can be seen that higher concentrations of overall immobilized biocide require less contact time to completely inactivate bacterial cells. For instance, a statistically significant ($p < 0.05$) six-log reduction (until no CFU count) was observed after a five-min contact time when using the overall biocide concentration of 3000 mg/L. The contact time required to obtain no CFU count was respectively 15 min and 30 min for 1000 mg/L and 500 mg/L immobilized biocide concentrations.

Furthermore, Figure 6a shows that the control particles (Al_2O_3-DA) had no impact on the culturability, since no decrease on CFU/mL was observed. Statistically significant differences ($p < 0.05$) were found between the biocidal particles (at all the tested concentrations) and the control particles, confirming the effective bactericidal activity of the BAC-loaded particles and the non-existing antimicrobial activity of the core–base materials. Although some studies report the antimicrobial activity of dopamine, such antimicrobial activity seemed to be related with the high concentration of DA tested. For example, Zhao et al. [43] reported that at 100 mg/mL of DA coating had an antimicrobial effect against *E. coli* cells, while at 30 times lower concentrations, Iqbal et al. [44] found that DA did not exhibit any antimicrobial activity against the same bacterium. Indeed, the concentration in the Al_2O_3-DA particles in this study is only 2 mg/mL of DA, which agrees with the idea that low concentrations of DA do not have an antimicrobial effect.

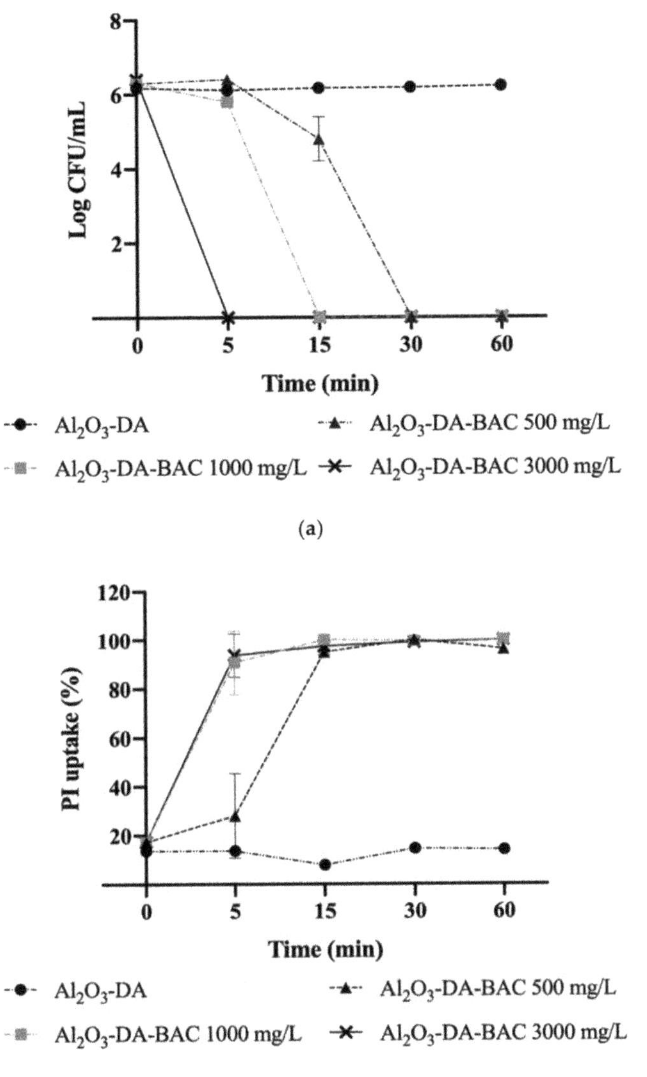

Figure 6. Logarithm of the colony forming units per milliliter (CFU/mL) (**a**) and propidium iodide (PI) uptake (%) (**b**) of *Escherichia coli* planktonic cells exposed to dopamine (DA)-functionalized particles over time during the first use. The overall tested concentrations were: 500, 1000 and 3000 mg/L. of Al_2O_3, and Al_2O_3-DA particles were used as controls. Error bars correspond to the standard deviation of the mean determined for two independent experiments.

2.3.2. Membrane Integrity

The effect of the functionalized particles over time on the membrane integrity (represented in terms of PI uptake %) of the bacterial cells was also studied for different biocide concentrations- see Figure 6b. As expected, the impact of the control particles (Al_2O_3-DA, before functionalization), on membrane integrity loss was negligible when compared to the particles containing the immobilized biocide ($p < 0.05$).

The results show that total loss of membrane integrity (100% PI uptake) was achieved with functionalized Al_2O_3-DA-BAC particles for all tested concentrations, although the lower biocide concentration (500 mg/L) took some more time to reach very high levels (above 90%) of cell membrane permeabilization. It could be said that increasing the overall immobilized biocide concentration decreases the contact time needed to fully permeabilize bacterial membrane and reduce culturability. This is in accordance to what is widely described in the literature concerning the concentration and contact time of free biocides [45,46]. Additionally, Pazos-Ortiz et al. [47] found that bacterial inhibition was dose-dependent, when studying the antimicrobial activity of silver nanoparticles. However, the results obtained in the present work, concerning a liquid biocide incorporated in inert particles (similar curves for 1000 and 3000 mg/L, with faster effects than for the lower concentration of 500 mg/L) leave some doubt about the possible existence of a critical concentration above which the biocide concentrations remain equally efficient.

The conclusions are also not so straightforward when comparing culturability and membrane permeabilization. Looking at Figure 6a,b, a similar level of membrane permeabilization is observed at five min for concentrations of 3000 mg/L and 1000 mg/L, but the CFU log reduction is much more pronounced for the higher concentration than for 1000 mg/L. These discrepancies suggest that the membrane damage caused by the antimicrobial agent was not enough to affect the bacterial growth on solid media, depending on the biocide concentration.

There are several studies in the literature reporting differences between these two methods, mostly related to the existence of more viable than culturable cells. One of such examples is the work of Ferreira, et al. [48], who observed that some cells with intact membranes (no PI uptake) were not culturable—the so called "viable but nonculturable" state. Nonetheless, it has also been reported by Rosenberg et al. [49] that PI-based viability tests can overestimate the number of dead cells, since they found that a dual-species 24-h biofilm presented 96% of PI-positive cells, although 68% of those cells were still metabolically active, and more than 80% of these cells were cultivable after harvesting. Furthermore, it has been shown that membranes from cells exposed to stress conditions (starvation, heat, biocide, etc.), tend to become more permeable to PI, resulting in mistakenly marking viable cells as dead [50–52]. Interestingly, it was also found that if such cells were re-incubated for a certain period after being exposed to the stress conditions, they could recover and repair the membrane damage [50].

The results discussed herein suggest that membrane integrity assays do not correlate with the ability of bacteria to grow on solid medium. Therefore, they confirm the importance of complementing the "viability" assays methods with a cultivation method [53] or to incubate the cells prior to the membrane integrity staining procedure [50].

2.4. *Effect of Particle Reuse on Their Antimicrobial Activity*

After being used once (first use), the same particles were reused and their antimicrobial activity regarding culturability (Figure 7a) and PI uptake (Figure 7b) was re-evaluated. In general, the particles lost some performance upon reuse, i.e., reusing the particles at 3000 and 1000 mg/L required more time to reach the maximum efficiency in terms of CFU count and % of PI uptake. This issue is addressed in Section 2.5, where the mechanisms of particles' antimicrobial activity are discussed.

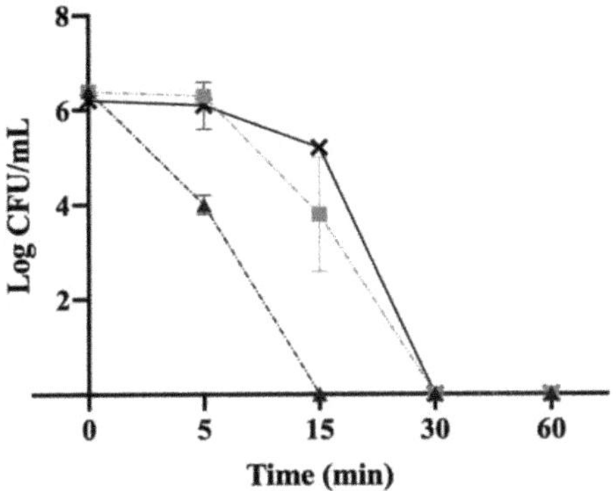

(a)

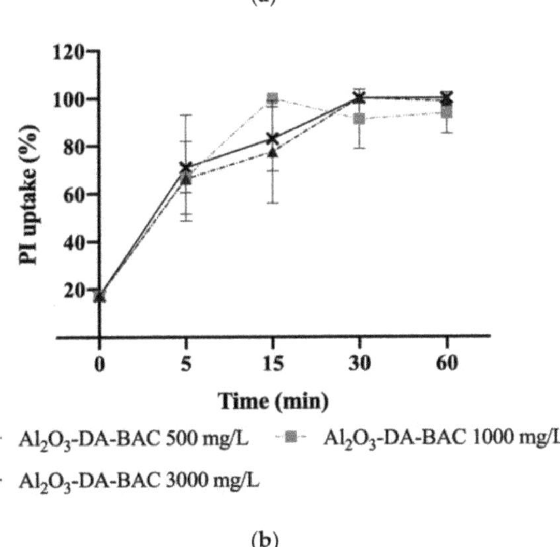

(b)

Figure 7. Logarithm of the colony forming units per milliliter (CFU/mL) (**a**) and propidium iodide (PI) uptake (%) (**b**) of *Escherichia coli* planktonic cells exposed to functionalized particles over time after reuse. The overall tested concentrations were: 500, 1000 and 3000 mg/L. Error bars correspond to the standard deviation of the mean determined for two independent experiments.

However, there is a specific case that stands out: Al_2O_3-DA-BAC at 500 mg/L. In this case, at first sight there would be an improvement of the particles' performance on reuse, since the contact time until total inactivation seems to decrease from 30 min on the first use to 15 min upon reuse. This can be related to the large period between sampling time points. Most probably, the total inactivation might occur at min 16 or 17. This would not be detected in the present tests, but the actual performance of the Al_2O_3-DA-BAC at 500 mg/L would be the same in the first use and in reuse.

Similar to what was observed in the first use, the PI uptake percentages (Figure 7b) do not match with the CFU results (Figure 7a). For example, regarding Al_2O_3-DA-BAC at 500 mg/L, although the cells lost their culturability at 15 min, 22% of them did not present damaged membranes. Possibly, this may be related to the fact that the cells have reached the "viable but non-culturable" state [49].

Due to the differences observed after reuse, a TG analysis of the particles after being reused was performed (see Figure 8). It is important to refer that the samples used were particles from the 3000-mg/L experiments.

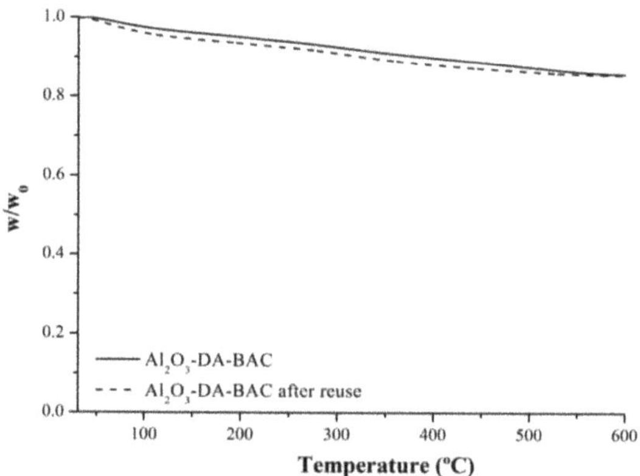

Figure 8. Thermogravimetric analysis (TG) of Al_2O_3-DA-BAC particles before the antimicrobial experiments and after being reused.

From Figure 8, it is possible to see that the curves are very similar. The amount of BAC in the samples after use is still in the order of 3% (w/w), meaning that there was no significant release of BAC. The results from the zone of inhibition seem to indicate a higher release, but it is known that the biocide/antibiotic concentrations in agar are higher than in liquid medium, causing higher death rates [54]. The zone of inhibition is only a qualitative method, while TG gives quantitative results. Furthermore, in the inhibition zone experiments the particles are in contact with bacteria for 24 h, which is much longer than the contact time of the killing experiments (one hour). Gathering all data, it is possible to conclude that there is a small and slow release of biocide from the particles.

2.5. *Hypothesis for the Underlying Mechanism of Action of Immobilized Biocide*

The results observed after reuse can be related to the following main question: "what are the mechanisms of antimicrobial action of biocide-loaded particles?" There are some hypotheses, namely: (i) bacteria are inactivated by the free biocide resulting from the slow release to the bulk solution, (ii) direct contact with the immobilized biocide (also known as "contact killing") is responsible for the antimicrobial action and (iii) there is a combination of both mechanisms of action stated above.

If the first hypothesis was behind the antimicrobial action of the particles, the reuse should have a similar behavior. Since the amount of biocide immobilized per unit mass of particle is very high, it would be expected that upon reuse there would still remain a considerable amount of biocide to be released. Additionally, the hypothesis that all or most of the immobilized biocide could be released during the first use is not plausible, otherwise the killing kinetics for the 3000 and 1000 mg/L concentrations would be faster.

The second hypothesis is that bacteria can be killed by contact with the BAC molecules immobilized onto the particles surface [54,55]. Several authors have been developing contact-active materials, which consist of coating a surface with a biocidal layer [56–58]. Synthetic QACs are by far the most used molecules for this purpose [59]. The mechanism of antimicrobial action of QACs seems to be due to electrostatic interactions between the positively charged QAC molecules and the negatively charged phospholipids present in the cytoplasmatic membrane of bacteria. Furthermore, the hydrophobic alkyl chain of QACs punches the membrane, leading to membrane disruption, cellular content leakage and microbial death [59–61]. Moreover, for the interaction between BAC molecules and bacteria to occur, the particles need to be in close contact with bacteria. Of course, the BAC molecules inside the pores will not be readily available for bacteria [62], since the particles are mesoporous (pore diameters between 2 and 50 nm [63]) and bacteria (due to their bigger size) cannot enter the pores. In the work of He et al. [56], they found that bacteria are dead upon contact with the immobilized QACs. However, it is known that these antimicrobial moieties can be blocked by dead adherent bacteria [56,64]. This mechanism of contact killing and consequent biocide blocking by dead bacteria may apply to the results observed in our study. Attraction between the particles, which display positive surface charge at pH around 7 (because the isoelectric point is pH = 8.2; see Figure 4), and the negatively charged bacterial surfaces is most probable to occur. It is therefore plausible that some dead bacteria from the 1st use remained attached to the particles' surface, which would explain the performance decrease upon reuse.

The third hypothesis assumes that the mechanism of action is a combination of the first and the second hypotheses. In fact, as it was already discussed, the first theory alone is not plausible, but the results from the zone of inhibition show some biocide leaching, which cannot be neglected and could contribute to some (although low) antimicrobial action of the biocide in solution.

Work is now being carried out in order to fully understand these interactions between functionalized particles and bacteria. The implementation of additional analytical methodologies to deeply characterize the mechanism of action of the immobilized BAC and compare it with the mechanism of free biocide seems also to be a key step for further understanding of the results. Finally, assuming that particles lose some of their performance, several cycles of reuse assays should be performed in order to characterize their lifetime effectiveness. If so, for real-field applications, a particle regeneration (cleaning) step would be periodically needed, implying the simultaneous operation of two or more alternating parallel particle bed systems.

3. Materials and Methods

3.1. Reagents

Dopamine (3-Hydroxytyramine Hydrochloride) was obtained from Tokyo Chemical Industry Co., Ltd. (Tokyo, Japan). Benzalkonium chloride ≥95 was obtained from Sigma-Aldrich (Steinheim, Germany). Tris base was obtained from ChemCruz™ Biochemicals (Dallas, Tx, USA).

3.2. Preparation and Functionalization of Particles

Alumina oxide pellets (Al_2O_3, Saint-Gobain, Stow, MA, USA) with approximately 3 mm were selected as start material. Pellets activation was accomplished in a heat treatment at 600 °C under air atmosphere for 6 h.

3.2.1. Initial Functionalization of Alumina Oxide Particles with DA

In this work, an amino-functionalization strategy was tested using dopamine (DA) as intermediate for BAC immobilization.

Dopamine hydrochloride (2 mg/mL) was dissolved in Tris buffer solution (pH = 8.5, 50 mM). The particles (5 g) were placed in contact with 50 mL of dopamine solution and left in agitation for 24 h. These Al_2O_3-DA particles were washed with distilled water to neutral pH, dried for 24 h in the oven at 120 °C and stored in a desiccator.

3.2.2. Immobilization of Biocide on the Surface of Alumina Oxide Particles

Samples of Al_2O_3-DA particles (2 g) were added to 100 mL of 5% (*w*/*v*) QAC solution and allowed to interact for 72 h at room temperature and with agitation speed of 140 rpm. The resulting particles (Al_2O_3-DA-BAC) were washed with distilled water to neutral pH, dried for 24 h in the oven at 80 °C and stored in a desiccator.

3.3. *Particles Characterization*

3.3.1. Thermal Stability of the Particles

Thermogravimetric analyses (TG) were performed to determine the particles content in terms of functional groups, after each immobilization step. It has been performed on a Mettler-Toledo TGA/DSC 1 STAR system. The thermal stability of samples was evaluated by heating the different samples up to 600 °C at 3 °C/min under air atmosphere and monitoring their weight loss.

3.3.2. Textural Characterization of the Particles

The textural characterization of the materials was based on the N_2 physisorption adsorption–desorption isotherms, determined at −196 °C with Quantachrome Autosorb IQ2. The surface area (S_{BET}) was calculated from nitrogen adsorption isotherms using the Brunauer, Emmet, and Teller (BET) equation [65]. Pore size distributions were obtained from the desorption branch of the isotherm using the Barrett, Joyner and Halenda (BJH) method [66].

3.3.3. Structural Properties of the Particles

Phase composition of the samples was analyzed by means of powder X-ray diffraction using a PanAnalytical X Pert PRO diffractometer set at 45 kV and 40 mA, using Cu Kα radiation (λ = 1.541874 Å) and a PIXcel detector. Data were collected using Bragg-Brentano configuration in the 2θ range of 20 to 80 ° with a scan speed of 0.01 °/s

3.3.4. Determination of Particle Surface Charge

The Zeta potential measurements were carried out using Dynamic Light Scattering by a Nano Particle Analyzer SZ-100 (Horiba Scientific, Longjumeau cedex, France). The particles (in powder) were placed in ultrapure water and the pH was adjusted from 2 to 11 by using an appropriate amount of 0.1-M NaOH or 0.1-M HCl. After being sonicated for 30 min, the particles surface charge was determined. At least 5 measurements were performed for each sample and the measurements repeated in two different occasions.

3.3.5. Fourier-Transform Infrared Spectroscopy (FTIR)

Spectroscopic measurements were performed with a Vertex 80v FTIR spectrometer (Bruker Optics) using Attenuated total reflectance (ATR) mode. The samples between 400 and 4000 cm^{-1} were analyzed with 64 scans averaging 4 cm^{-1} and two independent experiments were performed for each sample.

3.4. *Particles Antimicrobial Activity*

3.4.1. Microorganism

Bacterial suspensions of *Escherichia coli* CECT 434 were obtained from an overnight growth at 37 °C and under agitation (120 rpm) in R2A broth medium with the following composition (per liter): 0.5 g glucose (CHEM-LAB, Zedelgem, Belgium), 0.5 g peptone (Oxoid, Hampshire, England, UK), 0.1 g $MgSO_4.7H_2O$ (Merck, Darmstadt, Germany), 0.5 g casein hydrolysate (Oxoid, Hampshire, England, UK), 0.3 g sodium pyruvate (Fluka, Steinheim, Germany), 0.5 g starch (Sigma-Aldrich, Steinheim, Germany), 0.5 g yeast extract (Merck, Darmstadt, Germany) and 0.4 g $K_2HO_4P{\cdot}3H_2O$ (Applichem Panreac, Darmstadt, Germany).

3.4.2. Antimicrobial Activity of the Particles against *E. coli*

The antimicrobial activity of the particles was tested against *E. coli* by using the Kirby–Bauer disk diffusion assay. The overnight culture was diluted to achieve 10^6 CFU/mL. Then, 100 µL of the test suspension was swabbed on Plate Count Agar (PCA) plates. Afterwards, the particles were placed on the center of the inoculated agar plates. Antibacterial activity was assessed by measuring the diameter of the inhibition zone (mm)—a circular area around the particle where bacteria has not grown—on the surface of the plates. Two independent experiments with triplicates were performed for each case.

3.4.3. Effect of Particle Functionalization and Biocide Concentration on Antimicrobial Activity

The antimicrobial activity of the biocidal particles was tested against *E. coli*. An overnight culture was centrifuged at 4000 rpm for 12 min. The pellets were resuspended in phosphate buffered saline (PBS) buffer solution, washed twice and resuspended in the same solution. The bacterial count was then adjusted to 10^6 CFU/mL. The particles were incubated with the bacterial cells for 5, 15, 30 and 60 min. As stated in the Introduction section, the aim of this work is to gather critical information on the properties and performance of the biocidal particles to proceed with the design of an antimicrobial particle bed-reactor. Therefore, by varying the number of particles, three different ratios of mass of particles per volume of solution were tested, corresponding to the following overall immobilized biocide concentrations (mass of immobilized biocide per total volume of solution): 500 mg/L, 1000 mg/L and 3000 mg/L (Table 3). These ratios were established attending to the bed-reactor dimensions, the particles characteristics (dimensions and density) and considering the wall effect approach [12]. The overall tested concentrations also considered the volume of solution (which was kept the same in all assays: 25 mL) and the amount of immobilized BAC per unit mass of particles (3%). The amount of biocide incorporated onto the particles' surface was determined by TG, by calculating the mass loss that occurred at the BAC decomposition temperature range.

The biocidal activity was evaluated in terms of culturability and membrane integrity due to propidium iodide (PI) uptake. After being exposed to the particles, the bacterial suspension was successively diluted in PBS and culturability was assessed on Plate Count Agar (PCA) plates using the drop plate method [67]. Thereafter, plates were incubated at 37 °C for 24 h and the colonies enumerated. The culturability results are expressed as logarithm of the Colony Forming Units per milliliter (CFU/mL), with bacterial detection limit being 10^2 CFU/mL (2 log). For membrane integrity assessment, the LIVE/DEAD® *Bac*light™ kit (Invitrogen) was used. This kit is composed by two nucleic acids stains: SYTO9™ and PI. The last compound only penetrates into cells with damaged membranes, staining them in red. SYTO9™ crosses all bacterial membranes, staining the cells with green color [48,68]. To implement the method, cells were diluted 1:10 in PBS and 1 mL aliquot was filtered on a 0.2-μm Nucleopore® (Whatman) black polycarbonate membrane and stained with 250 μL of SYTO9™ and 250 μL of PI. The stain reagents reacted for 7 min in the dark at room temperature. After that, the excess reagents were filtered, and the membrane mounted on a slide with *Bac*Light mounting oil. The microscope, software, and the emission and excitation filters used were the same as described by Ferreira et al. [48]. The results were represented in terms of PI uptake percentage. Two independent experiments were performed for each case.

3.4.4. Effect of Reuse on Particles' Antimicrobial Activity

After the first use (1st use), the functionalized particles were washed with 1 L of distilled water and dried at 120 °C for 24 h. Then, the particles were reused once, and their antimicrobial activity was re-evaluated as described in Section 3.4.3. Hereafter, this second use of the particles will be named—reuse.

3.5. Statistical Analysis

Data obtained for culturability and membrane integrity were analyzed using the statistical software GraphPad Prism 8.0 (GraphPad Software, Inc., San Diego, California USA). The influence of biocide concentration and reuse were evaluated using a two-way ANOVA with Tukey's multiple comparison test. Differences were considered relevant if $p < 0.05$. All the statistical analyses were performed using the GraphPad Prism 8 software (GraphPad Software, Suite, San Diego, California, USA).

4. Conclusions

Alumina oxide particles were successfully functionalized with the antimicrobial agent benzalkonium chloride (BAC). To the best of our knowledge, this was the first work that accomplished the immobilization of biocides on the surface of macroscopic metal oxide particles, using DA (dopamine) as a precursor for biocide fixation. The functionalized particles (Al_2O_3-DA-BAC) showed a consistent antimicrobial effect against *E. coli*, which increased with the overall immobilized biocide concentration. In general, there was some decrease in the particles' performance on reuse, corresponding to an increase in the contact time required to reach similar results to the 1st use (for the same conditions). Even so, a high antimicrobial activity was still achieved upon reuse, indicating the feasibility of the Al_2O_3-DA-BAC particles for water disinfection, as a final polishing step before Reverse Osmosis treatment. The total cost of the core particles and the functionalization methods here described is relatively low when compared to the cost of, for example, silver nanoparticles used in lab scale experiments and are expected to be substantially reduced when going to an industrial scale production. In order to optimize the performance of the functionalized particles, the mechanism of action of the immobilized biocide needs to be studied in detail to understand how the immobilized biocide interacts with bacteria, in comparison with the free biocide mechanisms. Additionally, the possible interactions between functionalized particles and dead bacteria (or dead cell components) should be further investigated and characterized.

Author Contributions: Formal analysis, A.C.B., A.P. and L.F.M.; methodology, A.C.B. and J.P.S.S.; supervision, A.P., L.F.M. and J.P.S.S.; writing—original draft, A.C.B. and A.P. and writing—review and editing, A.C.B., A.P., L.F.M. and J.P.S.S. All authors have read and agreed to the published version of the manuscript.

Funding: This work was financially supported by: (i) Base Funding—UIDB/00511/2020 of the Laboratory for Process Engineering, Environment, Biotechnology and Energy—LEPABE—funded by national funds through the FCT/MCTES (PIDDAC); (ii) Project INN-INDIGO/0001/2014- POMACEA—Affordable technology for mitigation of membrane (bio) fouling through oPtimizatiOn of pretreatMent And ClEAIng methods, funded by the ERA-NET-European Research Area Networks program through the FCT, by national funds.; and (iii) Project pBio4.0—Preventing Biofouling in Membrane Systems, with the reference POCI-01-0247-FEDER-033298, co-funded by the European Regional Development Fund (ERDF), through the Operational Programme for Competitiveness and Internationalization (COMPETE 2020), under the PORTUGAL 2020 Partnership Agreement. A.C.B acknowledges the receipt of a Ph.D. grant from FCT (SFRH/BD/146028/2019).

Institutional Review Board Statement: Not applicable.

Informed Consent Statement: Not applicable.

Data Availability Statement: Not applicable.

Conflicts of Interest: The authors declare no conflict of interest.

References

1. Boretti, A.; Rosa, L. Reassessing the projections of the World Water Development Report. *npj Clean Water* **2019**, *2*, 1–6. [CrossRef]
2. Vrouwenvelder, J.; Van der Kooij, D. Diagnosis, prediction and prevention of biofouling of NF and RO membranes. *Desalination* **2001**, *139*, 65–71. [CrossRef]
3. Majamaa, K.; Johnson, J.E.; Bertheas, U. Three steps to control biofouling in reverse osmosis systems. *Desalination Water Treat.* **2012**, *42*, 107–116. [CrossRef]
4. Vrouwenvelder, J.; Beyer, F.; Dahmani, K.; Hasan, N.; Galjaard, G.; Kruithof, J.; Van Loosdrecht, M. Phosphate limitation to control biofouling. *Water Res.* **2010**, *44*, 3454–3466. [CrossRef] [PubMed]
5. Araújo, P.A.; Mergulhão, F.; Melo, L.; Simões, M. The ability of an antimicrobial agent to penetrate a biofilm is not correlated with its killing or removal efficiency. *Biofouling* **2014**, *30*, 675–683. [CrossRef]
6. Xiong, Y.; Liu, Y. Biological control of microbial attachment: A promising alternative for mitigating membrane biofouling. *Appl Microbiol Biotechnol* **2010**, *86*, 825–837. [CrossRef]
7. Singer, A.C.; Shaw, H.; Rhodes, V.; Hart, A. Review of antimicrobial resistance in the environment and its relevance to environmental regulators. *Front. Microbiol.* **2016**, *7*, 1728. [CrossRef]

8. Khairnar, S.; Shinde, S.; Shrivastava, V. A Short Review on the Improvement of Antimicrobial Activity by Metal and Nonmetal Doping in Nanoscale Antimicrobial Materials. *J. Nanomedicine Biotherapeutic Discov.* **2019**, *9*, 163.
9. Qu, X.; Alvarez, P.J.; Li, Q. Applications of nanotechnology in water and wastewater treatment. *Water Res.* **2013**, *47*, 3931–3946. [CrossRef]
10. Akbar, A.; Sadiq, M.B.; Ali, I.; Muhammad, N.; Rehman, Z.; Khan, M.N.; Muhammad, J.; Khan, S.A.; Rehman, F.U.; Anal, A.K. Synthesis and antimicrobial activity of zinc oxide nanoparticles against foodborne pathogens *Salmonella typhimurium* and *Staphylococcus aureus*. *Biocatal. Agric. Biotechnol.* **2019**, *17*, 36–42. [CrossRef]
11. Azizi-Lalabadi, M.; Ehsani, A.; Divband, B.; Alizadeh-Sani, M. Antimicrobial activity of Titanium dioxide and Zinc oxide nanoparticles supported in 4A zeolite and evaluation the morphological characteristic. *Scientific reports* **2019**, *9*, 1–10. [CrossRef] [PubMed]
12. Benyahia, F.; O'Neill, K. Enhanced voidage correlations for packed beds of various particle shapes and sizes. *Part. Sci. Technol.* **2005**, *23*, 169–177. [CrossRef]
13. Emami-Karvani, Z.; Chehrazi, P. Antibacterial activity of ZnO nanoparticle on Gram-positive and Gram-negative bacteria. *Afr. J. Microbiol. Res.* **2012**, *5*, 1368–1373.
14. Jaswal, V.S.; Chaudhary, A.; Thakur, P.; Sharma, D.; Arora, A.K.; Khanna, R.; Tuli, H.S. Chapter 9: ZnO nanoparticle with promising antimicrobial and antiproliferation synergistic properties. In *Comprehensive Analytical Chemistry: Engineered Nanomaterials and Phytonanotechnology: Challenges for Plant Sustainability*; Verma, S.K., Das, A.K., Eds.; Elsevier: Amsterdam, The Netherlands, 2019; Volume 87, pp. 251–262.
15. Kim, J.S.; Kuk, E.; Yu, K.N.; Kim, J.-H.; Park, S.J.; Lee, H.J.; Kim, S.H.; Park, Y.K.; Park, Y.H.; Hwang, C.-Y.; et al. Antimicrobial effects of silver nanoparticles. *Nanomedicine* **2007**, *3*, 95–101. [CrossRef]
16. Le Ouay, B.; Stellacci, F. Antibacterial activity of silver nanoparticles: A surface science insight. *Nano Today* **2015**, *10*, 339–354. [CrossRef]
17. Tang, S.; Zheng, J. Antibacterial Activity of Silver Nanoparticles: Structural Effects. *Adv. Healthc. Mater.* **2018**, *7*, 1701503. [CrossRef]
18. Yan, X.; He, B.; Liu, L.; Qu, G.; Shi, J.; Hu, L.; Jiang, G. Antibacterial mechanism of silver nanoparticles in *Pseudomonas aeruginosa*: Proteomics approach. *Metallomics* **2018**, *10*, 557–564. [CrossRef]
19. Asri, L.A.T.W.; Crismaru, M.; Roest, S.; Chen, Y.; Ivashenko, O.; Rudolf, P.; Tiller, J.C.; van der Mei, H.C.; Loontjens, T.J.A.; Busscher, H.J. A Shape-Adaptive, Antibacterial-Coating of Immobilized Quaternary-Ammonium Compounds Tethered on Hyperbranched Polyurea and its Mechanism of Action. *Adv. Funct. Mater.* **2014**, *24*, 346–355. [CrossRef]
20. Avelelas, F.; Martins, R.; Oliveira, T.; Maia, F.; Malheiro, E.; Soares, A.M.; Loureiro, S.; Tedim, J. Efficacy and ecotoxicity of novel anti-fouling nanomaterials in target and non-target marine species. *Mar. Biotechnol.* **2017**, *19*, 164–174. [CrossRef]
21. Chan, A.C.; Bravo Cadena, M.; Townley, H.E.; Fricker, M.D.; Thompson, I.P. Effective delivery of volatile biocides employing mesoporous silicates for treating biofilms. *J. R. Soc. Interface* **2017**, *14*, 20160650. [CrossRef] [PubMed]
22. Xue, Y.; Xiao, H.; Zhang, Y. Antimicrobial polymeric materials with quaternary ammonium and phosphonium salts. *Int, J. Mol. Sci* **2015**, *16*, 3626–3655. [CrossRef]
23. Wang, J.-L.; Alasonati, E.; Tharaud, M.; Gelabert, A.; Fisicaro, P.; Benedetti, M.F. Flow and fate of silver nanoparticles in small French catchments under different land-uses: The first one-year study. *Water Res.* **2020**, *176*, 115722. [CrossRef]
24. Akhtar, K.; Khan, S.A.; Khan, S.B.; Asiri, S.M. Nanomaterials and environmental remediation: A fundamental overview. In *Nanomaterials for Environmental Applications and their Fascinating Attributes*; Khan, S.B., Asiri, S.M., Akhtar, K., Eds.; Bentham Science Publishers: Sharjah, UAE, 2018; Volume 2, pp. 26–27.
25. Zhang, C.; Cui, F.; Zeng, G.-m.; Jiang, M.; Yang, Z.-z.; Yu, Z.-g.; Zhu, M.-y.; Shen, L.-q. Quaternary ammonium compounds (QACs): A review on occurrence, fate and toxicity in the environment. *Sci Total Environ.* **2015**, *518*, 352–362. [CrossRef]
26. Gerba, C.P. Quaternary ammonium biocides: Efficacy in application. *Appl. Environ. Microbiol.* **2015**, *81*, 464–469. [CrossRef] [PubMed]
27. García, M.R.; Cabo, M.L. Optimization of *E. coli* inactivation by benzalkonium chloride reveals the importance of quantifying the inoculum effect on chemical disinfection. *Front. Microbiol.* **2018**, *9*, 1259. [CrossRef] [PubMed]
28. Gao, D.; Feng, J.; Ma, J.; Lü, B.; Jia, X. Zinc oxide sol-containing diallylmethyl alkyl quaternary ammonium salt synthesized by sol–gel process: Characterization and properties. *J. Text. Inst.* **2015**, *106*, 593–600. [CrossRef]
29. Muhammad, S.; Siddiq, M.; Niazi, J.H.; Qureshi, A. Role of quaternary ammonium compound immobilized metallic graphene oxide in PMMA/PEG membrane for antibacterial, antifouling and selective gas permeability properties. *Polym. Bull.* **2018**, *75*, 5695–5712. [CrossRef]
30. Choi, J.; Han, Y.; Park, S.; Park, J.; Kim, H. Pore characteristics and hydrothermal stability of mesoporous silica: Role of oleic acid. *J. Nanomater.* **2014**, *2014*, 86. [CrossRef]
31. She, X.; Chen, L.; Li, C.; He, C.; He, L.; Kong, L. Functionalization of hollow mesoporous silica nanoparticles for improved 5-FU loading. *J. Nanomater.* **2015**, *16*, 108. [CrossRef]
32. Zhao, W.; Chen, H.; Li, Y.; Li, L.; Lang, M.; Shi, J. Uniform rattle-type hollow magnetic mesoporous spheres as drug delivery carriers and their sustained-release property. *Adv. Func Mater.* **2008**, *18*, 2780–2788. [CrossRef]
33. Liu, J. Multilayered PEI-based Films for CO2 Adsorption and Diffusion. MSc diss., University of Akron. 2013. Available online: http://rave.ohiolink.edu/etdc/view?acc_num=akron1367839488 (accessed on 15 October 2020).

34. Ribena, D. Dopamine modification of interfaces between polymers and metals. Ph.D. Thesis, Technische Universiteit Eindhoven, Eindhoven, The Netherlands, 2012.
35. Padhye, L.; Luzinova, Y.; Cho, M.; Mizaikoff, B.; Kim, J.-H.; Huang, C.-H. PolyDADMAC and dimethylamine as precursors of N-nitrosodimethylamine during ozonation: Reaction kinetics and mechanisms. *Environ. Sci. Technol.* **2011**, *45*, 4353–4359. [CrossRef]
36. Park, S.-H.; Wei, S.; Mizaikoff, B.; Taylor, A.E.; Favero, C.; Huang, C.-H. Degradation of amine-based water treatment polymers during chloramination as N-nitrosodimethylamine (NDMA) precursors. *Environ. Sci. Technol.* **2009**, *43*, 1360–1366. [CrossRef] [PubMed]
37. Hassan, K.; Elbagoury, M. Antimicrobial activity of some biocides against microorganisms isolated from a shared student kitchen. *Rasayan, J. Chem.* **2018**, *11*, 238–244.
38. Drugeon, H.B.; Juvin, M.-E.; Caillon, J.; Courtieu, A.-L. Assessment of formulas for calculating critical concentration by the agar diffusion method. *Antimicrob. Agents Chemother.* **1987**, *31*, 870–875. [CrossRef] [PubMed]
39. Cooper, K.E. Chapter 1: The Theory of Antibiotic Inhibition Zones. In *Analytical Microbiology*; Kavanagh, F., Ed.; Academic Press: London, UK, 1972; Volume 2, pp. 13–30.
40. Lee, D.; Cohen, R.E.; Rubner, M.F. Antibacterial properties of Ag nanoparticle loaded multilayers and formation of magnetically directed antibacterial microparticles. *Langmuir* **2005**, *21*, 9651–9659. [CrossRef] [PubMed]
41. Green, J.-B.D.; Fulghum, T.; Nordhaus, M.A. A review of immobilized antimicrobial agents and methods for testing. *Biointerphases* **2011**, *6*, MR13–MR28. [CrossRef]
42. Murray, P.R. The clinician and the microbiology laboratory. In *Mandell, Douglas, and Bennett's Principles and Practice of Infectious Diseases*; Elsevier: Amsterdam, The Netherlands, 2015; Volume 1, pp. 191–223.
43. Zhao, P.; Li, J.; Wang, Y.; Jiang, H. Broad-spectrum antimicrobial activity of the reactive compounds generated in vitro by *Manduca sexta* phenoloxidase. *Insect Biochemi Mol. Biol* **2007**, *37*, 952–959. [CrossRef]
44. Iqbal, Z.; Lai, E.P.; Avis, T.J. Antimicrobial effect of polydopamine coating on *Escherichia coli*. *J. Mater. Chem.* **2012**, *22*, 21608–21612. [CrossRef]
45. Capita, R.; Vicente-Velasco, M.; Rodríguez-Melcón, C.; García-Fernández, C.; Carballo, J.; Alonso-Calleja, C. Effect of low doses of biocides on the antimicrobial resistance and the biofilms of *Cronobacter sakazakii* and *Yersinia enterocolitica*. *Sci. rep.* **2019**, *9*, 1–12. [CrossRef]
46. Grant, D.; Bott, T. Biocide dosing strategies for biofilm control. *Heat Transf. Eng.* **2005**, *26*, 44–50. [CrossRef]
47. Pazos-Ortiz, E.; Roque-Ruiz, J.H.; Hinojos-Márquez, E.A.; López-Esparza, J.; Donohué-Cornejo, A.; Cuevas-González, J.C.; Espinosa-Cristóbal, L.F.; Reyes-López, S.Y. Dose-dependent antimicrobial activity of silver nanoparticles on polycaprolactone fibers against gram-positive and gram-negative bacteria. *J. Nanomater.* **2017**, *6*, 1–9. [CrossRef]
48. Ferreira, C.; Pereira, A.; Pereira, M.; Melo, L.; Simões, M. Physiological changes induced by the quaternary ammonium compound benzyldimethyldodecylammonium chloride on *Pseudomonas fluorescens*. *J. Antimicrob. Chemother.* **2011**, *66*, 1036–1043. [CrossRef]
49. Rosenberg, M.; Azevedo, N.F.; Ivask, A. Propidium iodide staining underestimates viability of adherent bacterial cells. *Sci. Rep.* **2019**, *9*, 1–12. [CrossRef]
50. Davey, H.M.; Hexley, P. Red but not dead? Membranes of stressed *Saccharomyces cerevisiae* are permeable to propidium iodide. *Environ. Microbiol.* **2011**, *13*, 163–171. [CrossRef] [PubMed]
51. Kirchhoff, C.; Cypionka, H. Propidium ion enters viable cells with high membrane potential during live-dead staining. *J. Microbiol. Methods* **2017**, *142*, 79–82. [CrossRef] [PubMed]
52. Shi, L.; Günther, S.; Hübschmann, T.; Wick, L.Y.; Harms, H.; Müller, S. Limits of propidium iodide as a cell viability indicator for environmental bacteria. *Cytometry Part. A* **2007**, *71*, 592–598. [CrossRef]
53. Emerson, J.B.; Adams, R.I.; Román, C.M.B.; Brooks, B.; Coil, D.A.; Dahlhausen, K.; Ganz, H.H.; Hartmann, E.M.; Hsu, T.; Justice, N.B. Schrödinger's microbes: Tools for distinguishing the living from the dead in microbial ecosystems. *Microbiome* **2017**, *5*, 86. [CrossRef]
54. Lu, G.; Wu, D.; Fu, R. Studies on the synthesis and antibacterial activities of polymeric quaternary ammonium salts from dimethylaminoethyl methacrylate. *React. Funct. Polym.* **2007**, *67*, 355–366. [CrossRef]
55. Owusu-Adom, K.; Guymon, C.A. Photopolymerization kinetics of poly (acrylate)–clay composites using polymerizable surfactants. *Polymer* **2008**, *49*, 2636–2643. [CrossRef]
56. He, W.; Zhang, Y.; Li, J.; Gao, Y.; Luo, F.; Tan, H.; Wang, K.; Fu, Q. A novel surface structure consisting of contact-active antibacterial upper-layer and antifouling sub-layer derived from gemini quaternary ammonium salt polyurethanes. *Sci. Rep.* **2016**, *6*, 1–9. [CrossRef]
57. Ho, C.H.; Tobis, J.; Sprich, C.; Thomann, R.; Tiller, J.C. Nanoseparated polymeric networks with multiple antimicrobial properties. *Adv. Mater.* **2004**, *16*, 957–961. [CrossRef]
58. Sui, Y.; Gao, X.; Wang, Z.; Gao, C. Antifouling and antibacterial improvement of surface-functionalized poly (vinylidene fluoride) membrane prepared via dihydroxyphenylalanine-initiated atom transfer radical graft polymerizations. *J. Membr. Sci.* **2012**, *394*, 107–119. [CrossRef]
59. Andrade del Olmo, J.; Ruiz Rubio, L.; Saez Martinez, L.; Perez-Alvarez, V.; Vilas Vilela, J.L. Antibacterial coatings for improving the performance of biomaterials. *Coatings* **2020**, *10*, 139. [CrossRef]

60. Bieser, A.M.; Tiller, J.C. Mechanistic considerations on contact-active antimicrobial surfaces with controlled functional group densities. *Macromol. Biosci.* **2011**, *11*, 526–534. [CrossRef]
61. Hoque, J.; Akkapeddi, P.; Yadav, V.; Manjunath, G.B.; Uppu, D.S.; Konai, M.M.; Yarlagadda, V.; Sanyal, K.; Haldar, J. Broad spectrum antibacterial and antifungal polymeric paint materials: Synthesis, structure–activity relationship, and membrane-active mode of action. *ACS Appl. Mater. Interfaces* **2015**, *7*, 1804–1815. [CrossRef]
62. Ricardo, S.I.d.C. Antimicrobial strategies to prevent catheters-associated medical infections. Ph.D. Thesis, Universidade de Lisboa, Lisbon, Portugal, 2017.
63. Amgoth, C.; Phan, C.; Banavoth, M.; Rompivalasa, S.; Tang, G. Polymer Properties: Functionalization and Surface Modified Nanoparticles. In *Role of Novel Drug Delivery Vehicles in Nanobiomedicine*; IntechOpen: London, UK, 2019.
64. Tuson, H.H.; Weibel, D.B. Bacteria–surface interactions. *Soft. matter.* **2013**, *9*, 4368–4380. [CrossRef]
65. Brunauer, S.; Emmett, P.H.; Teller, E. Adsorption of Gases in Multimolecular Layers. *J. Am. Chem. Soc.* **1938**, *60*, 309–319. [CrossRef]
66. Barrett, E.P.; Joyner, L.G.; Halenda, P.P. The Determination of Pore Volume and Area Distributions in Porous Substances. I. Computations from Nitrogen Isotherms. *J. Am. Chem. Soc.* **1951**, *73*, 373–380. [CrossRef]
67. Reed, R.; Reed, G. "Drop plate" method of counting viable bacteria. *Can. J. Res.* **1948**, *26*, 317–326. [CrossRef]
68. Simoes, M.; Pereira, M.O.; Vieira, M. Validation of respirometry as a short-term method to assess the efficacy of biocides. *Biofouling* **2005**, *21*, 9–17. [CrossRef]

Article

Antibacterial Property of Cellulose Acetate Composite Materials Reinforced with Aluminum Nitride

Thefye P. M. Sunthar [1,2,*], Francesco Boschetto [1,3], Hoan Ngoc Doan [4], Taigi Honma [1], Kenji Kinashi [4], Tetsuya Adachi [3], Elia Marin [1,3], Wenliang Zhu [1] and Giuseppe Pezzotti [1,2,5,6]

1 Ceramic Physics Laboratory, Kyoto Institute of Technology, Sakyo-ku, Matsugasaki, Kyoto 606-8585, Japan; boschetto-cesc@kit.ac.jp (F.B.); m0672026@edu.kit.ac.jp (T.H.); elia-marin@kit.ac.jp (E.M.); wlzhu@kit.ac.jp (W.Z.); pezzotti@kit.ac.jp (G.P.)
2 Department of Immunology, Graduate School of Medical Science, Kyoto Prefectural University of Medicine Kamigyo-ku, 465 Kajii-cho, Kawaramachi dori, Kyoto 602-0841, Japan
3 Department of Dental Medicine, Graduate School of Medical Science, Kyoto Prefectural University of Medicine, Kamigyo-ku, Kyoto 602-8566, Japan; t-adachi@koto.kpu-m.ac.jp
4 Faculty of Material Science and Engineering, Kyoto Institute of Technology, Sakyo-ku, Matsugasaki, Kyoto 606-8585, Japan; ngochoandoan@gmail.com (H.N.D.); kinashi@kit.ac.jp (K.K.)
5 The Center for Advanced Medical Engineering and Informatics, Osaka University, Yamadaoka, Suita, Osaka 565-0871, Japan
6 Department of Orthopedic Surgery, Tokyo Medical University, 6-7-1 Nishi-Shinjuku, Shinjuku-ku, Tokyo 565-0871, Japan
* Correspondence: thefyeprasad@rocketmail.com

Abstract: Cellulose acetate (CA) is a synthetic compound that is derived from the acetylation of cellulose. CA is well known as it has been used for many commercial products such as textiles, plastic films, and cigarette filters. In this research, antibacterial CA composites were produced by addition of aluminum nitride (AlN) at different weight percentage, from 0 wt. % to 20 wt. %. The surface characterization was performed using laser microscope, Raman and FTIR spectroscopy. The mechanical and thermal properties of the composite were analyzed. Although the mechanical strength tended to decrease as the concentration of AlN increased and needed to be optimized, the melting temperature (T_m) and glass transition temperature (T_g) showed a shift toward higher values as the AlN concentration increased leading to an improvement in thermal properties. AlN additions in weight percentages >10 wt. % led to appreciable antibacterial properties against *S. epidermidis* and *E. coli* bacteria. Antibacterial CA/AlN composites with higher thermal stability have potential applications as alternative materials for plastic packaging in the food industry.

Keywords: aluminum nitride; composite; antibacterial; mechanical; thermal; cellulose acetate

Citation: Sunthar, T.P.M.; Boschetto, F.; Doan, H.N.; Honma, T.; Kinashi, K.; Adachi, T.; Marin, E.; Zhu, W.; Pezzotti, G. Antibacterial Property of Cellulose Acetate Composite Materials Reinforced with Aluminum Nitride. *Antibiotics* **2021**, *10*, 1292. https://doi.org/10.3390/antibiotics10111292

Academic Editors: Luís Melo and Andreia S. Azevedo

Received: 21 September 2021
Accepted: 20 October 2021
Published: 22 October 2021

1. Introduction

Food provides nutritional support for any organism. However, most people around the world are not aware of the dangers related to the lack of proper food hygiene. According to a survey from World Health Organization (WHO), 600 million cases of foodborne diseases and 420,000 deaths were recorded worldwide in the year 2015. The common bacteria that responsible for the foodborne illness are *Norovirus*, *Listeria*, *Campylobacter jejuni*, *Salmonella*, *Staphylococcus aureus*, *Escherichia coli* [1,2].

Food contamination may happen during harvesting, processing, packaging and distribution [2]. However, packaging plays an important role to protect food from being affected by various kinds of contaminants and preserve the products from biological, chemical and physical changes while storage or during preparation. In many so-called "advanced countries", the food quality is a very important factor—they used to reject it even if there was a small change in the smell or appearance [1].

In literature, many different techniques were applied to improve the antimicrobial properties of packaging systems [2]. Some of the methods are based on the reinforcement of volatile and nonvolatile antimicrobial compounds directly into the polymers like the application of sachet or pads that contain volatile antimicrobial compounds, applying antimicrobial coating compounds on the surface of the polymers, ions or covalent linkages of immobilization of antimicrobial agent into the polymers [3–5]. In other cases, chitosan was directly used without any modification as an antimicrobial coating film [3–6]. After consideration of the previous research, this research is aimed to implement the new and easy idea to produce the antibacterial composite material using easily degradable cellulose acetate material as the main substrate and reinforced with AlN which could replace the usage of plastic material.

Cellulose acetate (CA) is a synthetic compound which is derived from the acetylation of cellulose. Cellulose is a natural polymer obtained from plant fibers, with the chemical formula $C_6H_7O_2$ $(OH)_3$ [7]. However, pure cellulose has a complex structure that cannot be easily modulated by using heat or solvents. The acetylation process causes the hydrogen in the hydroxyl groups is replaced by acetyl groups (CH_3CO) which turns it into cellulose acetate. CA is easier to dissolved in certain solvents like acetone or can be melted under heat and molded into solid objects, spun into fibers or cast as a film [8].

Aluminum nitride (AlN) is well known for its excellent thermal conductivity, high coefficient of thermal expansion, high electrical resistivity and high dielectric strength [9]. Besides the wide use of AlN in semiconductors, it showed antimicrobial properties such as those of Si_3N_4 [10]. This property is due to the reactivity of the AlN with water which produces ammonia (NH_3) and ammonium ions (NH_4^+) that eventually kill the bacteria [10].

This research focuses on producing alternative food packaging materials, plastic tablecloths or toys by reinforcing CA with AlN. The CA/AlN composites can be produced with a lower cost and non-complicated processes. In addition, it also can show both mechanical and thermal stability with enhanced antibacterial property.

Possible future developments include degradable CA/AlN composites which might contribute to the reduction of environmentally harmful plastic waste.

2. Materials and Methods

2.1. Samples Preparation

The samples were prepared as showed in Figure 1. Cellulose acetate powder was crushed using pestle and mortar to obtain fined powder. Then the powder was mixed with acetone until completely dissolved and 0.05 mL of pure triacetin were mixed to the solution, as a plasticizer. Once the solution cleared, agglomerated AlN powder were poured into the glass vial and mixed by stirring until a homogenous mixture is obtained. The liquid is then poured inside a flat mold obtained by using polyimide tape on the Teflon plate. Then the film was casted using glass plate to fill the entire row to form a layer. The casting process were repeated 5 times to obtain a desire thickness. The film was left overnight to dry at room temperature. The film was removed from the plate after the acetone evaporated and forming a layer of the CA/AlN composites. It was then heated at 60 °C in a vacuum for overnight to remove the remaining solvent inside the film. Five different samples were produced with 0 wt. %, 5 wt. %, 10 wt. %, 15 wt. % and 20 wt. % of AlN, respectively, (i.e., CA content = 100 wt. % − AlN wt. %).

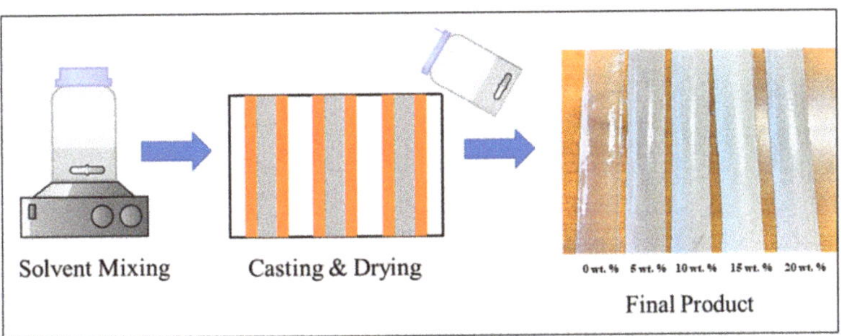

Figure 1. Experimental procedure of CA/AlN composites.

2.2. *Sample Characterization*

2.2.1. Laser Microscopy

The surface morphology of the sample was analyzed with the aid of a confocal scanning laser microscope (Laser Microscope 3D and Profile measurements, Keyence, VK × 200 Series, Osaka, Japan) equipped with a numerical aperture between 0.30 and 0.95. It has the x-y stage and autofocus function for z range. The micrographs were collected ranging from 10× to 150× to evaluate macroscopic and microscopic roughness of the samples. 25 images randomly selected from the surface of the map and the micrographs were then analyzed using Keyence Color 3D Laser Microscope VK-X100/X200 series VK Analyzer software (Keyence, Osaka, Japan).

2.2.2. Fourier Transformed Infrared Spectroscopy

Fourier Transformed Infra-Red spectroscopy (FTIR) spectra were collected at room temperature using an FTIR spectrometer (ATR-FTIR, FTIR-4700 with ATR PRO ONE equipped with a diamond prism, Jasco Co., Tokyo, Japan) with a Michelson 28-degree interferometer with corner-cube mirrors with a range between 250,000 and 5 cm^{-1}. The aperture size was 200 × 200 μm^2 the acquisition time was fixed to 30 s. The instrument was operated using (Spectra Manager, JASCO, Tokyo, Japan) software. A total of three samples of each type were scanned from 400 and 4000 cm^{-1} at 5 different locations. The spectra were analyzed with (OriginLab Co., Northampton, MA, USA, and LabSpec, Horiba/Jobin-Yvon, Kyoto, Japan) software.

2.2.3. Raman Spectroscopy

Raman spectra of the samples were collected with the aid of triple monochromator (T-64000, Jobin-Ivon/Horiba Group, Kyoto, Japan) equipped with a charge coupled device (CCD) detector. The excitation source used is 532 nm Nd:YVO4 diode pumped solid-state laser (SOC JUNO, Showa Optronics Co. Ltd., Tokyo, Japan). In total, 25 randomly picked locations were investigated with spectrograph center wavelength 2500 cm^{-1}, grating 300 gr/mm, exposure time 4 s and average of 3. The resulting spectra were averaged. Raman spectral acquisition and pre-processing of raw data such as baseline subtraction, smoothing, normalization and fitting were acquired utilizing commercially available software (LabSpec, Horiba/Jobin-Yvon, Kyoto, Japan and Origin 8.5, OriginLab Co., Northampton, MA, USA).

2.3. *Mechanical and Thermal Properties*

2.3.1. Mechanical Properties

Tensile mechanical testing was conducted using a MCT 2150 Desktop Tensile-Compression Tester (AND Discover Precision, Tokyo, Japan) using a 500 N load cell at a strain rate of 50 mm min^{-1}. For testing, CA/AlN composite samples were cut using a standardized dumbbell shaped tensile sample cutter with an overall length of 35 mm, gauge length of 10 mm, distance between shoulders 12 mm, grip section 4.5 mm, width of grip section

6 mm, reduced section 12 mm. In total, 5 samples were tested with each concentration (0, 5, 10, 15, and 20 wt. %). The result was analyzed using Excel and OriginLab (Co., Northampton, MA, USA).

2.3.2. Thermal Properties

The thermal properties of the CA composite were investigated using a differential scanning calorimeter (DSC) (TA Q200, TA Instruments Japan Inc., Tokyo, Japan) with a heating/cool/heat cycle program. The sample was heated at 10 °C min^{-1} from −30 to 300 °C and cooled at 5 °C min^{-1} under a nitrogen atmosphere with a gas flow rate of 50 $\mu L \cdot h^{-1}$. Each sample was measured three times.

2.4. *In Vitro Testing*

The antibacterial analysis was conducted at the Kyoto Prefectural University of Medicine. Gram-positive bacteria, *Staphylococcus epidermidis* (14990™ ATCC® purchased from American Type Culture Collection (ATCC)) and Gram-negative bacteria, *Escherichia coli* (*E. coli*, ATCC® 25922™), were cultured using a brain heart infusion (BHI) liquid medium. The initial 1.8×10^{10} CFU/mL was subsequently diluted to 1.8×10^{8} CFU/mL using a phosphate-buffered saline (PBS, NACALAI TESQUE.INC, Kyoto, Japan) solution to mimic ion blood concentrations. The samples with dimensions of 1 cm × 1 cm were sterilized prior to the experiment using a UV sterilizer for 24 h. Then the samples were incubated at 37 °C for 12 and 24 h.

2.4.1. Microbial Viability Assay (WST)

WST is well known technique to measure the bacterial metabolism by calorimetric detection. In this experiment, the WST-8 kit (Microbial Viability Assay Kit-WST, Dojindo, Kumamoto, Japan) was used as a calorimetric indicator which releases a water-soluble formazan dye upon reduction in the presence of electron mediator. The amount of the formazan dye generated is linearly related to the number of living microorganisms. The solution is subjected to microplate readers (EMax, Molecular Devices, Sunnyvale, CA, USA) upon collecting the OD value related to living cells. Three samples were used to calculate the average values.

2.4.2. Crystal Violet Assay, Laser Microscopy Scanning, Viability Staining and ImageJ Analysis

A 0.5% crystal violet solution was used to determine the biofilm formation on the sample's surface. The PBS washed samples were placed into a 12-well plate filled with 500 µL crystal violet solution and incubated at room temperature for 20 min on shaker. The samples were then transferred to a new 6-well plate and washed four times gently with 5 mL PBS to excrete excess crystal violet. The plates were gently shaken to completely remove the residual dye. Then, the samples were measured using a laser microscope to measure the bacterial attachment and biofilm formation on the sample surface. The laser micrographs are then analyzed with ImageJ 1.50, to measure the area filled by bacteria. The analysis of biofilm density was measured by conversion of confocal images into red, green and blue colors. The background subtraction was applied to remove background noise. The images were then assembled into color channels and the integrated density of pixels was calculated. Correspondingly, the same samples were used to measure the OD of the samples with crystal violet stained which represent the amount of biofilm. The samples were transferred into a new 12-well plate filled with 95% ethanol and the plate was incubated at room temperature on a shaker at 270 rpm for 15 min. Then, 100 µL of the ethanol solution containing the crystal violet stained by the biofilms was transferred into the 96-well plate. The optical density at 595 nm was determined for total biomass quantification.

2.5. *Statistical Analysis*

The experimental data were statistically analyzed with Student's t-test (t-test) where $p < 0.05$ was considered statistically significant and marked with an asterisk (*) and not statistically significant data marked with (*ns*).

3. Results

3.1. *Surface Characterization*

3.1.1. Laser Microscopy

Figure 2 shows the laser micrographs images of five samples with different concentration of AlN ranging from 0 to 20 wt. %. The dispersion of AlN on the surface could be clearly observed as a function of concentration.

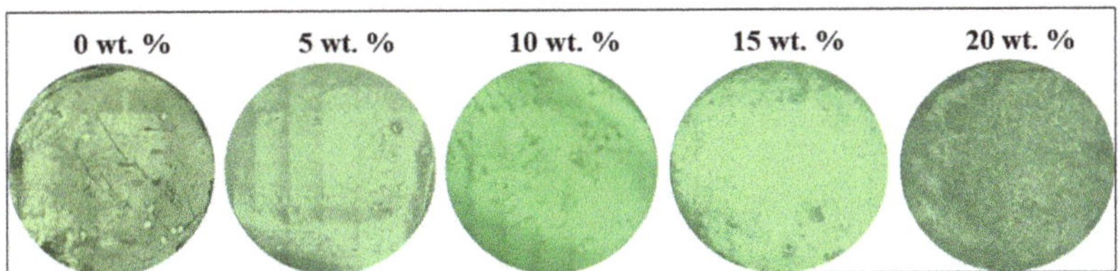

Figure 2. Laser microscope images of CA/AlN composites.

The white particles observed are the AlN. The surface roughness R_a increases as the amount of AlN increases. Figure 3 shows the quantification of surface roughness where a clear increasing trend with AlN concentration was observed. As shown, 20 wt. % shows the highest surface roughness and the whitish particle observed in image is AlN which increases the surface roughness of the material.

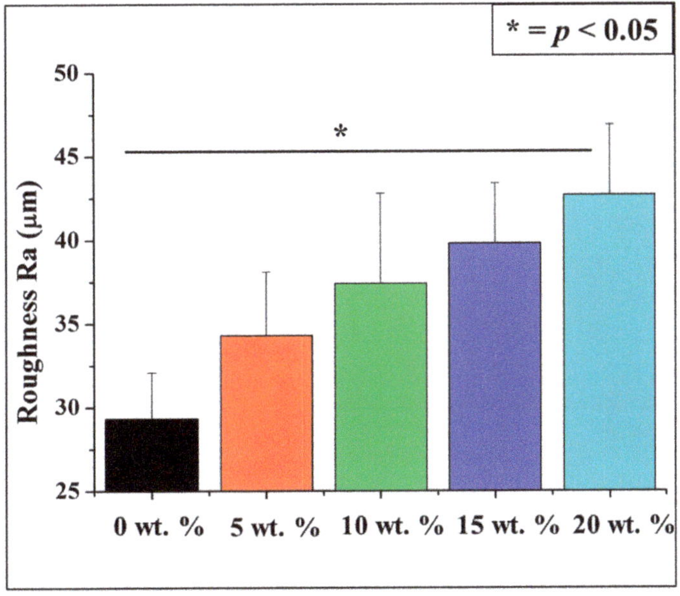

Figure 3. Surface microscopic roughness of CA/AlN composites. * = $p < 0.05$.

3.1.2. FTIR Spectroscopy

Figure 4 shows the FTIR spectra of the pure aluminum nitride, and the CA/AlN composites ranging from 0 wt. % to 20 wt. %. The composite material has a wide range spectrum from 500 to 3750 cm^{-1}. The FTIR spectra do not shows the presence of AlN as it can be clearly observed that none of the CA/AlN composites has the strong peak of AlN at 672 cm^{-1} (peak 1) which is assigned to E1(TO) [11–13]. Peak 2 to 10 represents cellulose acetate functional groups. The band at 1750 cm^{-1} (peak 7) was attributed to C=O from cellulose acetate and the band at 1428 cm^{-1} (peak 6) was assigned to CH_2 vibrations. The sharp absorption peaks at 1041 cm^{-1} (peak 3) and 1250 cm^{-1} (peak 4) were due to presence of C-O stretching [14,15]. The band at 912 cm^{-1} (peak 2) can be attributed to C-O stretching and CH_2 rocking vibrations. In addition, 1366 cm^{-1} (peak 5) was assigned to CH_3. The broad peak at and 2942 cm^{-1} (peak 9) and the peak at 3487 cm^{-1} (peak 10) attributed to C-H aromatic vibrations and O-H stretching of cellulose acetate [16,17], respectively. FTIR do not show any functional groups of AlN which are present in the CA/AlN composites. This might be because the strong signal of CA blocks the weak signals of AlN in the composite.

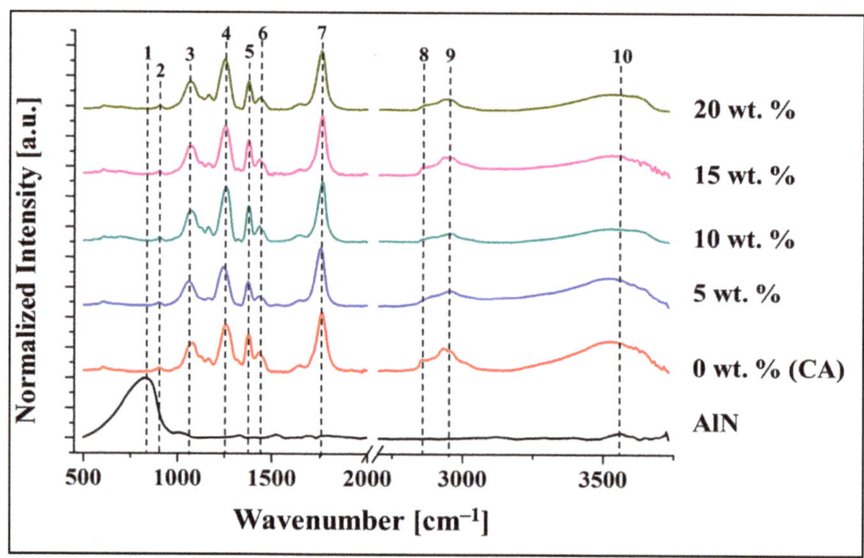

Figure 4. FTIR spectra of the CA/AlN composites.

3.1.3. Raman Spectroscopy

The functional groups of CA could by clearly observed by FTIR, while Raman spectroscopy was used to further investigate both the CA/AlN composites and AlN. The various spectra acquired were illustrated in Figure 5. The characteristics of Raman signal for cellulose was clearly observed at 2934 cm^{-1} (peak 6) which is assigned to C-H stretching and asymmetric stretching vibrations of the C-O-C glycosidic linkage. In addition, Raman signals at 1380 (peak 3), 1435 (peak 4) and 1754 cm^{-1} (peak 5) are attributed to C=O vibrations of the carbonyl group and asymmetric and symmetric vibrations of C-H bond which exist in the acetyl group from CA [17–19]. Besides, the presence of AlN could be clearly observed at 612 (peak 1) and 656 cm^{-1} (peak 2) which are associated with A_1 (TO) and E_2 (high) [20–22]. As AlN concentration increases the peak intensity also increases. The red spots are a marker for the presence of the AlN in the CA matrix, which is green. The intensity of the red spot increases in conjunction with the peak intensity, which can be explained with the increased presence of AlN. The AlN signals, which cannot be detected by FTIR, were clearly seen in the Raman spectra which proves the presence of AlN particles in the composite.

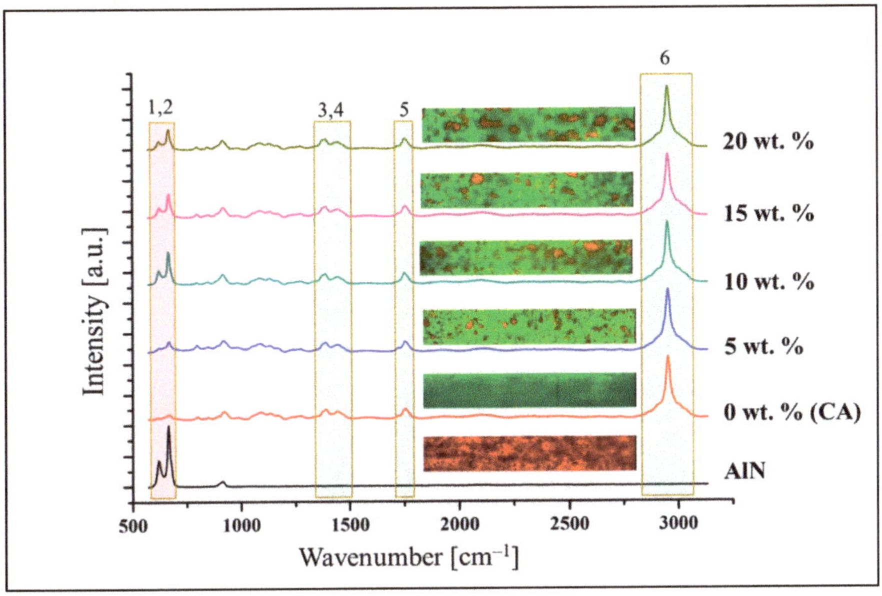

Figure 5. Raman average spectra of the CA/AlN composites.

3.2. Mechanical Property

Figure 6a,b illustrates the stress versus strain curves for the various CA/AlN composites. It provides the information on how the composite materials deforms with increasing force. The composites containing 5 wt. % of AlN has similar strength to the 0 wt. %. Mechanical strength was then reduced in the samples containing 10 wt. % to 20 wt. %. This proves that 5 wt. % reinforcement concentration has the highest strength among the composite samples.

Figure 6c shows the young modulus of the CA/AlN composites which were calculated using gradient of the stress versus strain curves. The Young's modulus of the samples containing 5 wt. % and 10 wt. % of AlN increases up to 1100 and 1221 MPa, respectively. However, when the further AlN were added, the Young's modulus of the 15 wt. % and 20 wt. % samples decrease to 989 and 892 MPa, respectively Therefore, it can be deduced that the stiffness of the samples increases with low AlN concentration (5 wt. % and 10 wt. %) and then reduce when the AlN reached to 15 and 20 wt. %.

In addition, the changes in toughness of the material were calculated by area un-der the graph as shown in Figure 6d. The toughness of the 0 wt. % sample was 2.7 MPa. When 5 wt. % of AlN was added to the composite, the toughness significantly increased, up to 4.0 MPa. which shows it is toughest sample. However, the toughness started to reduce when higher AlN concentrations were added to the composites. The calculations analysis was made as stated in the literature [18,23,24].

HDPE, PVC, and PVDC are the commercially available material which is used to produce cling film for food packaging, tablecloths and other plastic materials. Theoretically, the commercial HDPE, PVC and PVDC clings film has the tensile strength of $\geq$10, $\geq$15 and $\geq$60 MPa respectively. Therefore, the CA/AlN composites can be used as the food wrap or tablecloths since it has the tensile strength in the range of 57 to 41 MPa which meets the range of tensile strength of commercial product. As a summary, though the tensile strength reduces at 20 wt. % but it has the enough mechanical strength to be an alternative material when compared with the theoretical values of the commercial products. However, 20 wt. % showed high thermal and antibacterial property which enhance the composite material's efficiency.

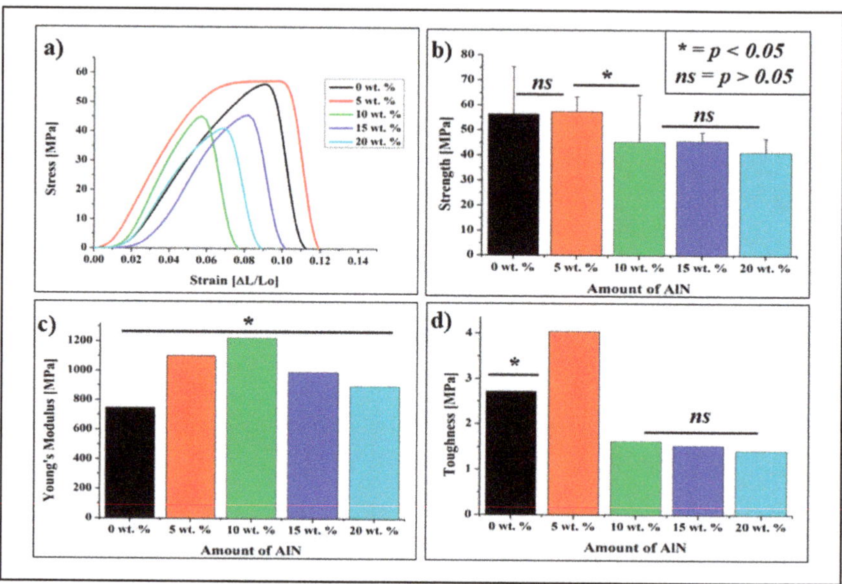

Figure 6. (**a**). Tensile stress-strain curves of the CA/AlN composites, (**b**) Max strength, (**c**) Young's modulus and (**d**) Toughness. * = $p < 0.05$, *ns* = $p > 0.05$.

3.3. Thermal Properties

The thermal property of the CA/AlN composites was analyzed by differential scanning calorimetry, DSC. The desorption temperature (T_d), glass transition temperature (T_g), melting temperature (T_m), and degree of crystallinity χ_c were measured. Figure 7 shows the first heating curve of DSC of cellulose acetate and CA/AlN composites. The first endothermic peak was clearly seen in all the samples. This peak appears due to the desorption of water. This phenomenon occurs due to the existence of residual moisture or low boiling point of solvents. T_d varies from control (0 wt. %) to 20 wt. % in the range from 71.82 °C to 79.97 °C. The differences of T_d values are due to their different ability of holding water of the polymeric matrices [14].

Figure 8a shows the second heating curve where the T_g and T_m could be identified. The T_g value were calculated by the maximum Tan delta peak [25–29]. The T_g values increases as the concentration of AlN increases as illustrated in Figure 8b. The T_g value of the CA improved with increasing the amount of AlN reinforcement.

The T_m temperature also shows similar trend where it increases as the amount of AlN increases. However, the increment seems to be small (about 7.35 °C from 0 wt. % to 20 wt. %). From Figure 8c, an endothermic peak appears in the first cycle close to the melting peak, but according to literature [26,29,30], the first heating curve is influenced by the thermal and mechanical history, and the second heating curve is conventionally used for the determination of material properties, such as meting point or glass transition temperature, under a given dynamic condition. For this reason, the melting point and glass transition temperature were investigated using the second heating curve [25–32].

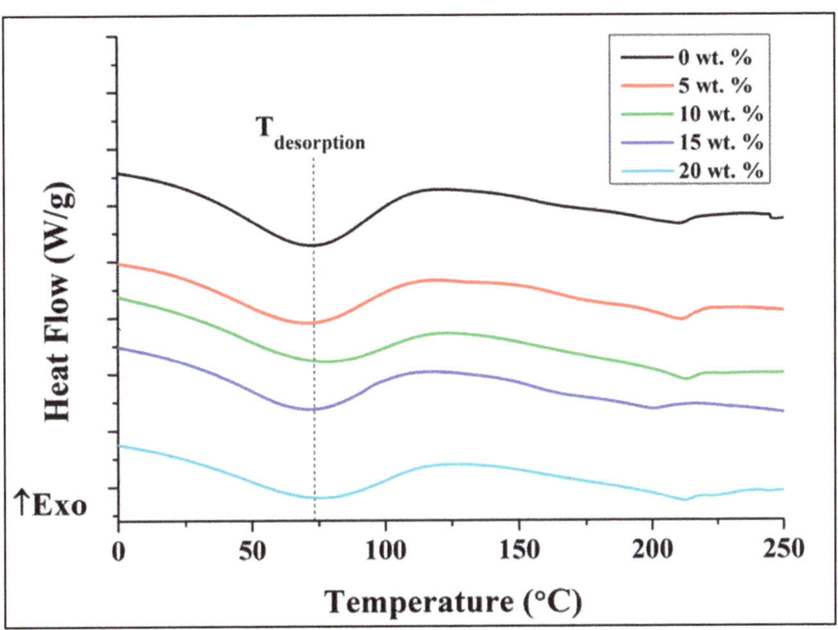

Figure 7. DSC first heating curves for the CA/AlN composites.

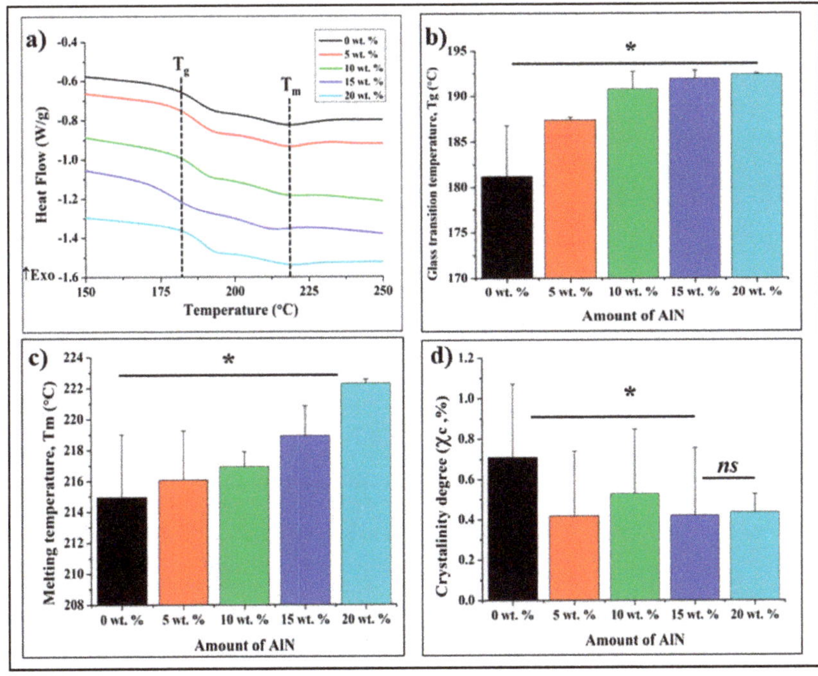

Figure 8. DSC curves for the CA/AlN composites (**a**), 2nd heating curve (**b**), Glass transition temperature, Tg, (**c**), Melting temperature and (**d**), crystallinity degree. * = $p < 0.05$, *ns* = $p > 0.05$.

Another parameter extrapolated from the DSC curves is the degree of crystallinity χ_c which is a fundamental property for properties of plastics. A higher degree of crystallization makes the material stiffer and stronger but also increases brittleness [26–31]. The crystallinity degree of CA/AlN composites was calculated as the ratio between the melting enthalpy of (ΔH_m) and the respective value for the totally crystalline material (ΔH_m^0) multiply with weight fraction of CA using the following formula:

$$\chi_c = \frac{\Delta H_m}{\omega \Delta H_m^0} \times 100 \tag{1}$$

where ΔH_m^0 = 58.8 J g^{-1} as stated by Cerquiera et al. [27].

Based on Figure 8d the degree of crystallinity reduces upon addition of AlN. However, the concentration of AlN does not affect the crystallinity degree much as the value was almost similar from 5 wt. % to 20 wt. %

3.4. *In Vitro Testing*

Microbial Viability Assay (WST)

Figure 9 demonstrates the antibacterial test conducted against the CA/AlN composites material with *S. epidermidis* using WST method. It is clearly observed that absorbance level increases until 15 wt. % and start to reduce at 20 wt. % for 12 h. Besides, the same samples were tested with the laser microscope, where the biofilm production was clearly seen on the surface as shown in Figure S1. Almost similar results were obtained for 12 h, the biofilm formation reduces slightly from the control sample and the antibacterial effect was not clearly observed. Whereas, for 24 h, the absorption shows the similar trend as 12 h, but at 20 wt. % shows a very lower absorption than control. The biofilm formation shows a very clear trend at 24 h where, at control it was observed the full area covered by patches of biofilm and it start to reduce as the concentration of AlN increases and at 20 wt. % and very low amount of biofilm was detected.

To reconfirm the antibacterial property, the samples were analyzed using Crystal violet (CV) staining method. The stained samples were analyzed with WST and Viability Staining, and ImageJ Analysis and the results were shown in Figure 9 The CV-stained samples show high absorbance at 12 h and start to reduce slightly at higher concentration especially at 15 and 20 wt. %, at 24 h the absorbance does not increase and was maintained as 0 wt. % sample but at 20 wt. % it decreases drastically. When compare with viability staining method, it shows the similar trend as the previous WST method where the biomass increases up to 15 wt. % and decreases at 20 wt. % for 12 h, whereas, at 24 h the biomass shows slight increment up to 15 wt. % and drastic drop at 20 wt. %.

For the *E. coli* bacteria, the WST measurement was not precise as the absorbance increases at 12 and 24 h even at 20 wt. % as shown in Figure 10. However, at 24 h, clear trend of decrement was observed especially at 20 wt. %. Comparing with the result of biofilm formation, the same trend was observed where the biofilm increases as the concentration increases. However, at 24 h the reduction of biofilm was clearly observed at 20 wt. % which in agreement with WST result. Although, laser micrographs images in Figure S2 (Supplementary Materials) shows the reduction of the biofilm formation as the concentration of AlN increases at 24 h.

In CV-stained experiment, the absorbance of *E. coli* increases as the concentration increases even at 20 wt. % as illustrated in Figure 11c,d. However, the absorbance shows a decrement at 20 wt. % compared with other concentrations. In the viability test, the biomass of *E. coli* reduces as the concentration increases but shows sudden spike at 20 wt. % for 12 h. However, at 24 h, a clear trend was observed where the biomass starts to reduce after 5 wt. % and lowest at 20 wt. %.

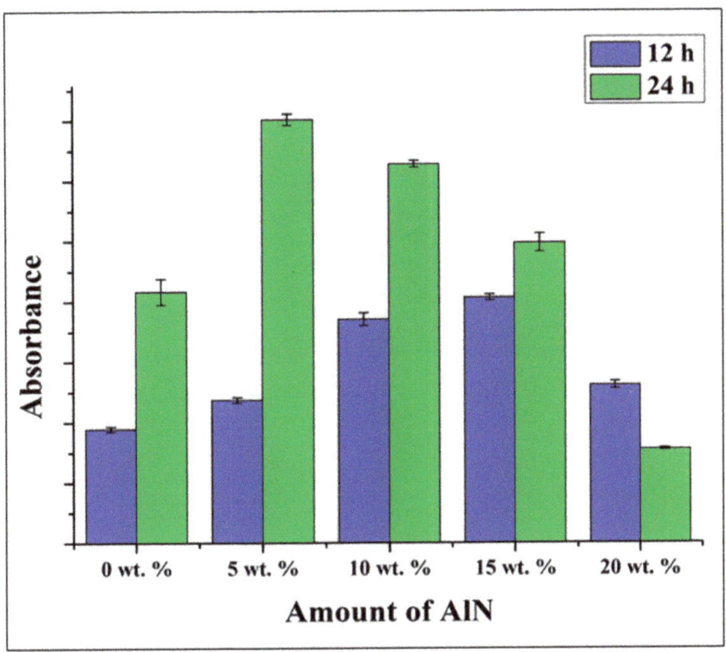

Figure 9. WST adsorption after 12 and 24 h of testing with *Staphylococcus epidermidis* on the CA/AlN composites, as a function of the fraction of AlN.

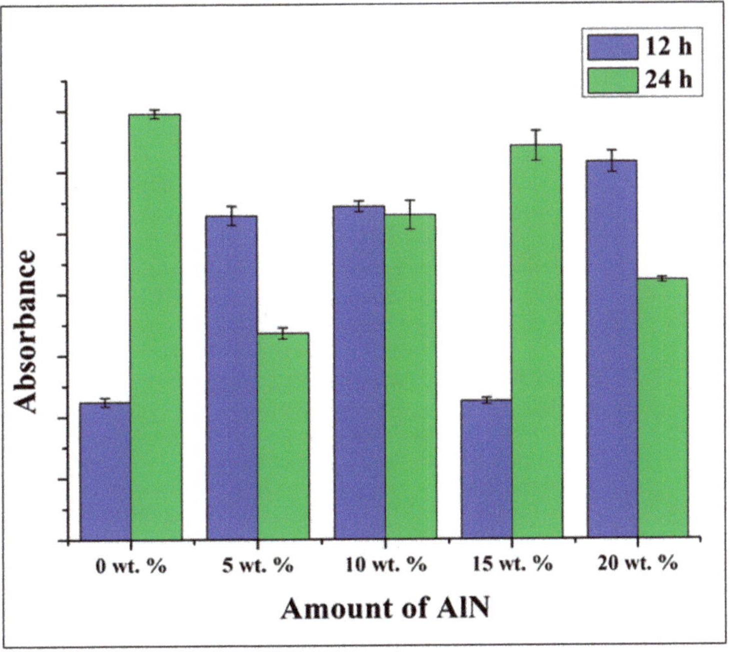

Figure 10. WST adsorption after 12 and 24 h of testing with *Escherichia coli* on the CA/AlN composites, as a function of the fraction of AlN.

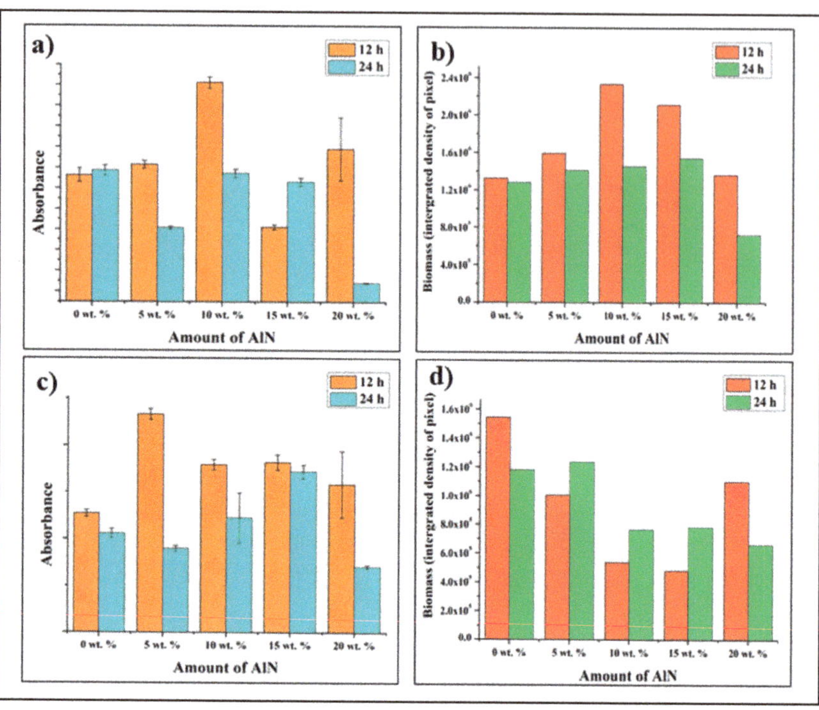

Figure 11. (**a**) *Staphylococcus epidermidis* biofilm formation after 12 and 24 h evaluated by quantitative measurement of crystal violet staining as an indicator of biomass accumulation on CA/AlN composites, (**b**) Quantitative comparison of accumulated biomass in (**a**), (**c**) *Escherichia coli* biofilm formation after 12 and 24 h evaluated by quantitative measurement of crystal violet staining as an indicator of biomass accumulation on CA/AlN composites, (**d**) Quantitative comparison of accumulated biomass in (**c**).

4. Discussion

The results of the laser microscope, FTIR and Raman spectroscopy clearly show the surface morphology and chemical composition of the CA/AlN composites. The surface roughness increases as a function of AlN concentration. Raman spectra clearly showed the presence of AlN particles which were hard to observe by FTIR. There was no observable alteration in the chemical bonds of both CA and AlN during the casting process which were clearly seen from FTIR and Raman spectroscopy.

The mechanical properties of the composites are affected by AlN concentration: the ultimate strength, the elongation and the toughness increase with a low concentration of AlN of 5 wt. %, then decreases with AlN concentration of 10 wt. % and above. The Young modulus of the composite also increases with AlN contents of 5 and 10 wt. %, then decreases when the AlN concentrations reach 15 and 20 wt. %. The mechanical behaviors of the CA/AlN composites could be explained by the phenomenon called "mechanical per-collation" [33]. Generally, in the polymer/filler composite system, the mechanical properties of the composite will increase until the filler concentration reaches a critical value, and then decreases with further filler content. The high concentration of the AlN particles could lead to the formation of agglomerates in the polymer matrix, affect the homogeneity of the CA/AlN composites, and causing lower mechanical properties.

The thermal properties were affected by the presence of reinforcing powders: the T_m and T_g values shows an increment. These increments could be attributed to the presence of AlN particles in the composite system. Due to the agglomeration of the filler particles, the mobility of the polymer chains is reduced. In order to mobilize the polymer chains, more energy is required, leading to the increase in the T_m and T_g values [34]. Whereas the degree of crystallinity of the composite material reduces when there is presence of AlN. However, the concentration does not affect the degree of crystallinity. The value was almost similar from 5 wt. % to 20 wt. %. Most importantly, this shows AlN can be a good reinforcing material.

The antibacterial properties were clearly shown in the Results chapter; however, it was difficult to observe the effect played by the increasing concentration of AlN which due to the releasing of active antibacterial component from the substrate. At higher concentrations, the antibacterial effect for both Gram-positive (*S. epidermidis)* and Gram-negative (*E. coli*) at 24 h was clearly observed. Secondly, the surface roughness at 20 wt. % might be the reason of the antibacterial effect which is expected to the liberation of ammonia (NH_3) from the surface of composite material into the aqueous solution upon exposing to water. The high surface roughness promotes more surface area which causes the bacteria to become exposed to the AlN and die. In total, 1 mole of AlN reacts with 3 mole of water and produces 1 mole of aluminum hydroxide and 1 mole of (NH_3). This reaction was considered as the overall hydrolysis reaction as given below [11,31,32,35–38]:

$$AlN + 3H_2O \rightarrow Al\,(OH)_3 + NH_3 \quad (2)$$

However, according to Bowen et al. [39] the hydrolysis of AlN in room temperature, the reaction can be classified into three processes as stated in following equations:

$$AlN + 2H_2O \rightarrow AlOOH_{(amorphous)} + NH_3 \quad (3)$$

$$NH_3 + H_2O \rightleftharpoons NH^+{}_4 + OH^- \quad (4)$$

$$AlOOH_{(amorphous)} + H_2O \rightarrow Al\,(OH)_3 \quad (5)$$

The release of ammonia reacts with the water to produce ammonium ions and hydroxide ions. The ammonium ions and ammonia release are responsible for the antibacterial property for the composite. There are few studies that have shown the antimicrobial property of ammonium salts. On the other side, the volatile ammonia (NH_3) gas release is expected to directly attack the structure of DNA of microorganisms [40–43]. Kleiner focused on the review of transportation of ammonia in bacteria and fungi which explains why bio membranes are highly permeable to free ammonia [39]. In another comprehensive study, the author claimed the ammonium ion, (NH_4^+) can only diffuse into the cytoplasmic space through ion channels and the tiny (NH_3) molecules can freely penetrate through the membrane [36,40–43]. Therefore, based on the previous studies and the results obtained, it can be speculated that the mechanism of antibacterial action is the elution of ammonia (NH_3) and ammonium ion (NH_4^+) during hydrolysis of AlN, as shown in Equations (1)–(3), diffuses into the bacterial cell and damages the DNA as well as causing cell lysis [35–43].

The amount of ammonium ion (NH_4^+) and ammonia (NH_3) released will determine the level of toxicity, thus a small preliminary experiment was conducted to measure the amount of NH_4^+ and NH_3 using a Quick Ammonia Meter AT-2000 instrument and the results were shown in Figure S3. The experiment was conducted at different time intervals of 2, 6, 12 and 24 h using the 20 wt. % of CA/AlN composite. The results demonstrate the release of NH_4^+ and NH_3 are very limited, and even at 24 h the maximum level of NH_4^+ and NH_3 are 0.44 mg/L and 0.28 mg/L respectively. Figure S4 shows the differences of the amount of NH_4^+ and NH_3 release by pure AlN powder and 20 wt. % of CA/AlN composite. The results clearly show the release of NH_4^+ and NH_3 is controlled by the CA substrate and the generation of aluminum hydroxide on the surface when exposed with water as shown in Equations (2)–(5) play a role to slow down the release. Therefore, the level of toxicity is much lower when compared to pure AlN powder. Secondly, the intended application of these CA/AlN composite is not for ingestion or biomedical devices:

it is intended to be used as tablecloth or food wraps which are not in direct contact with the human body environment.

5. Conclusions

The surface characterization of the composite material shows an increment in surface roughness due to presence of AlN, that could also be identified by Raman.

The mechanical strength of the composite was reduced at AlN fractions >10 wt. %. On the other hand, the Young's modulus showed an increase up to 10 wt. % and a decrease at a higher concentration. In addition, a clear decrement observed in toughness upon increasing concentration of AlN was observed.

The melting temperature (T_m) and glass transition temperature (T_g) increased with increasing AlN concentration, showing that the thermal properties of the CA/AlN composites were improved in the presence of AlN.

The CA/AlN composites showed antibacterial effects for both the Gram-positive and Gram-negative bacteria at higher concentration of AlN due to the reaction of AlN with water to produce ammonia (NH_3) and ammonium ions, which caused lysis by disruption of bacterial cell membrane.

In conclusion, this could be a promising material to replace plastic bags, food packaging or other plastic products thanks to improved antibacterial and thermal property.

Future research will focus on the degradability and stability of this composite material.

Supplementary Materials: The following are available online at https://www.mdpi.com/article/10.3390/antibiotics10111292/s1, Figure S1: Microscopic images of Staphylococcus epidermidis biofilm formation on the different CA/AlN composites after 12 h (a–e), 24 h (f–j), Figure S2: Microscopic images of Escherichia coli biofilm formation on the different CA/AlN composites after 12 h (a–e), 24 h (f–j), Figure S3: Quantification of Ammonium ion and Ammonia released during hydrolysis process of AlN at different time interval, Figure S4: Comparison of Ammonium ion and Ammonia released by AlN powder and 20 wt. % CA/AlN composite.

Author Contributions: Conceptualization, T.P.M.S.; methodology, T.A.; validation, T.A., G.P., E.M. and W.Z.; formal analysis, T.P.M.S., T.H. and H.N.D.; investigation, F.B., T.P.M.S. and T.H.; resources, T.P.M.S.; data curation, H.N.D., F.B., E.M. and W.Z.; writing—original draft preparation, T.P.M.S.; writing—review and editing, G.P., E.M., H.N.D., F.B. and W.Z.; visualization, K.K.; supervision, E.M., W.Z. and G.P.; project administration, T.P.M.S., E.M. and G.P. All authors have read and agreed to the published version of the manuscript.

Funding: This research project did not receive any specific grant from funding agencies in the public, commercial, or not-for-profit sectors.

Data Availability Statement: The data presented in this study are available on request from the corresponding author.

Conflicts of Interest: The authors declare no conflict of interest.

References

1. Sunthar, T.; Marin, E.; Boschetto, F.; Zanocco, M.; Sunahara, H.; Ramful, R.; Pezzotti, G. Antibacterial and ANTIFUNGAL properties of composite POLYETHYLENE Materials reinforced with Neem and turmeric. *Antibiotics* **2020**, *9*, 857. [CrossRef] [PubMed]
2. Kamala, K.; Kumar, V.P. Food products and food contamination. *Microb. Contam. Food Degrad.* **2018**, *9*, 1–19. [CrossRef]
3. Var, I. *Active Antimicrobial Food Packaging*; IntechOpen: London, UK, 2019. Available online: https://directory.doabooks.org/handle/20.500.12854/40079 (accessed on 19 October 2021).
4. Han, J.H. Antimicrobial food packaging. *Novel Food Packag. Tech.* **2003**, *8*, 50–70. [CrossRef]
5. Ahvenainen, R.; Ahvenainen, R. *Novel Food Packaging Techniques*; CRC Press: Boca Raton, FL, USA, 2003.
6. Khaneghah, A.M.; Hashemi, S.M.B.; Limbo, S. Antimicrobial agents and packaging systems in antimicrobial active food packaging: An overview of approaches and interactions. *Food Bioprod. Process.* **2018**, *111*, 1–19. [CrossRef]
7. Fischer, S.; Thümmler, K.; Volkert, B.; Hettrich, K.; Schmidt, I.; Fischer, K. Properties and applications of cellulose acetate. *Macromol. Symp.* **2008**, *262*, 89–96. [CrossRef]
8. Law, R.C. 5. applications of cellulose acetate 5.1 cellulose acetate in textile application. *Macromol. Symp.* **2004**, *208*, 255–266. [CrossRef]

9. Selvaduray, G.; Sheet, L. Aluminium nitride: Review of synthesis methods. *Mater. Sci. Tech.* **1993**, *9*, 463–473. [CrossRef]
10. Jovanović, G.D.; Klaus, A.S.; Nikšić, M.P. Antimicrobial activity of chitosan coatings and films against Listeria monocytogenes on black radish. *Rev. Argent. Microb.* **2016**, *48*, 128–136. [CrossRef] [PubMed]
11. Marin, E.; Boschetto, F.; Zanocco, M.; Honma, T.; Zhu, W.; Pezzotti, G. Explorative study on the antibacterial effects of 3D-printed PMMA/nitrides composites. *Mater. Design* **2021**, *206*, 109788. [CrossRef]
12. Molleja, J.G.; Gómez, B.J.; Abdallah, B.; Djouadi, M.A.; Feugeas, J.; Jouan, P.-Y. Study of AlN thin films deposited by DC magnetron sputtering: Effect of pressure on texture. *An. AFA* **2016**, *26*, 190–194. [CrossRef]
13. Ibáñez, J.; Hernández, S.; Alarcón-Lladó, E.; Cuscó, R.; Artús, L.; Novikov, S.V.; Foxon, C.T.; Calleja, E. Far-infrared transmission in gan, aln, AND algan thin films grown by molecular beam epitaxy. *J. Appl. Phys.* **2008**, *104*, 033544. [CrossRef]
14. Murphy, D.; de Pinho, M.N. An atr-ftir study of water in cellulose acetate Membranes prepared by phase inversion. *J. Membr. Sci.* **1995**, *106*, 245–257. [CrossRef]
15. Arkhangelsky, E.; Goren, U.; Gitis, V. Retention of organic matter by cellulose acetate membranes cleaned with hypochlorite. *Desalination* **2008**, *223*, 97–105. [CrossRef]
16. Sudiarti, T.; Wahyuningrum, D.; Bundjali, B.; Made Arcana, I. Mechanical strength and ionic conductivity of polymer electrolyte membranes prepared from cellulose acetate-lithium perchlorate. *IOP Conf. Ser. Mater. Sci. Eng.* **2017**, *223*, 012052. [CrossRef]
17. Dong, F.; Yan, M.; Jin, C.; Li, S. Characterization of Type-II Acetylated Cellulose Nanocrystals with Various Degree of Substitution and Its Compatibility in PLA Films. *Polymers* **2017**, *9*, 346. [CrossRef] [PubMed]
18. Sánchez-Márquez, J.A.; Fuentes-Ramírez, R.; Cano-Rodríguez, I.; Gamiño-Arroyo, Z.; Rubio-Rosas, E.; Kenny, J.M.; Rescignano, N. Membrane Made of Cellulose Acetate with Polyacrylic Acid Reinforced with Carbon Nanotubes and Its Applicability for Chromium Removal. *Int. J. Polym. Sci.* **2015**, *2015*, 1–12. [CrossRef] [PubMed]
19. Kubota, H.; Sakamoto, K.; Matsui, T. A confocal Raman Microscopic visualization of SMALL penetrants in cellulose acetate using A DEUTERIUM-LABELING TECHNIQUE. *Sci. Rep.* **2020**, *10*, 16426. [CrossRef] [PubMed]
20. Musa, I.; Qamhieh, N.; Said, K.; Mahmoud, S.T.; Alawadhi, H. Fabrication and characterization of Aluminum Nitride NANOPARTICLES by Rf magnetron sputtering and inert Gas CONDENSATION TECHNIQUE. *Coatings* **2020**, *10*, 411. [CrossRef]
21. Liu, L.; Liu, B.; Edgar, J.H.; Rajasingam, S.; Kuball, M. Raman characterization and stress analysis of AlN grown on SiC by sublimation. *J. Appl. Phys.* **2002**, *92*, 5183–5188. [CrossRef]
22. Li, X.; Zhou, C.; Jiang, G.; You, J. Raman analysis of aluminum nitride at high temperature. *Mater. Charact.* **2006**, *57*, 105–110. [CrossRef]
23. Yang, Z.-Y.; Wang, W.-J.; Shao, Z.-Q.; Zhu, H.-D.; Li, Y.-H.; Wang, F.-J. The transparency and mechanical properties of cellulose acetate nanocomposites using cellulose nanowhiskers as fillers. *Cellulose* **2013**, *20*, 159–168. [CrossRef]
24. Fogagnolo, J.B.; Robert, M.H.; Velasco, F.; Torralba, J.M. Aluminium matrix COMPOSITES reinforced WITH Si_3N_4, AlN And ZrB_2, produced by Conventional powder metallurgy and Mechanical Alloying. *KONA Powder Part. J.* **2004**, *22*, 143–150. [CrossRef]
25. Vo, P.; Doan, H.; Kinashi, K.; Sakai, W.; Tsutsumi, N.; Huynh, D. Centrifugally Spun Recycled PET: Processing and Characterization. *Polymers* **2018**, *10*, 680. [CrossRef] [PubMed]
26. de Freitas, R.R.M.; Senna, A.M.; Botaro, V.R. Influence of degree of substitution on thermal dynamic mechanical and physicochemical properties of cellulose acetate. *Ind. Crops Prod.* **2017**, *109*, 452–458. [CrossRef]
27. Barud, H.S.; de Araújo Júnior, A.M.; Santos, D.B.; de Assunção, R.M.N.; Meireles, C.S.; Cerqueira, D.A.; Rodrigues Filho, G.; Ribeiro, C.A.; Messaddeq, Y.; Ribeiro, S.J.L. Thermal behavior of cellulose acetate produced from homogeneous acetylation of bacterial cellulose. *Thermochim. Acta* **2008**, *471*, 61–69. [CrossRef]
28. Cerqueira, D.A.; Rodrigues Filho, G.; Assunção, R.M.N. A New Value for the Heat of Fusion of a Perfect Crystal of Cellulose Acetate. *Polym. Bull.* **2006**, *56*, 475–484. [CrossRef]
29. Erdmann, R.; Kabasci, S.; Heim, H.-P. Thermal Properties of Plasticized Cellulose Acetate and Its β-Relaxation Phenomenon. *Polymers* **2021**, *13*, 1356. [CrossRef] [PubMed]
30. Aoki, D.; Teramoto, Y.; Nishio, Y. Cellulose acetate/poly(methyl methacrylate) interpenetrating networks: Synthesis and estimation of thermal and mechanical properties. *Cellulose* **2011**, *18*, 1441–1454. [CrossRef]
31. Gutierrez, J.; Carrasco-Hernandez, S.; Barud, H.S.; Oliveira, R.L.; Carvalho, R.A.; Amaral, A.C.; Tercjak, A. Transparent nanostructured cellulose acetate films based on the self assembly OF PEO-b-PPO-b-PEO block COPOLYMER. *Carbohydr. Polym.* **2017**, *165*, 437–443. [CrossRef] [PubMed]
32. Iqhrammullah, M.; Marlina, M.; Khalil, H.P.; Kurniawan, K.H.; Suyanto, H.; Hedwig, R.; Karnadi, I.; Olaiya, N.G.; Abdullah, C.K.; Abdulmadjid, S.N. Characterization and performance evaluation of cellulose acetate–polyurethane film for lead ii ion removal. *Polymers* **2020**, *12*, 1317. [CrossRef] [PubMed]
33. Yang, C.; Dong, J.; Fang, Y.; Ma, L.; Zhao, X.; Zhang, Q. Preparation of novel low-κ polyimide fibers with simultaneously excellent mechanical properties, UV-resistance and surface activity using chemically bonded hyperbranched polysiloxane. *J. Mater. Chem. C* **2018**, *6*, 1229–1238. [CrossRef]
34. Liu, X.; Wang, T.; Chow, L.C.; Yang, M.; Mitchell, J.W. Effects of inorganic fillers on the thermal and mechanical properties of poly (lactic acid). *Int. J. Polym. Sci.* **2014**, *2014*, 827028. [CrossRef] [PubMed]
35. Kleiner, D. The transport of NH_3 and HN_4^+ across biological membranes. *Biochim. Biophys. Acta (BBA)-Rev. Bioenerg.* **1981**, *639*, 41–52. [CrossRef]

36. Pezzotti, G.; Asai, T.; Adachi, T.; Ohgitani, E.; Yamamoto, T.; Kanamura, N.; Boschetto, F.; Zhu, W.; Zanocco, M.; Marin, E.; et al. Antifungal activity of polymethyl methacrylate/Si3N4 composites against *Candida albicans*. *Acta Biomater.* **2021**, *126*, 259–276. [CrossRef] [PubMed]
37. Pezzotti, G.; Ohgitani, E.; Shin-Ya, M.; Adachi, T.; Marin, E.; Boschetto, F.; Zhu, W.; Mazda, O. Instantaneous "catch-and-kill" inactivation of SARS-CoV-2 by nitride ceramics. *Clin. Transl. Med.* **2020**, *10*, e212. [CrossRef] [PubMed]
38. Li, J.; Nakamura, M.; Shirai, T.; Matsumaru, K.; Ishizaki, C.; Ishizaki, K. Mechanism and Kinetics of Aluminum Nitride Powder Degradation in Moist Air. *J. Am. Ceram. Soc.* **2006**, *89*, 937–943. [CrossRef]
39. Bowen, P.; Highfield, J.G.; Mocellin, A.; Ring, T.A. ChemInform Abstract: Degradation of Aluminum Nitride Powder in an Aqueous Environment. *ChemInform* **1990**, *21*, 724–728. [CrossRef]
40. Valko, E.I.; DuBois, A.S. Correlation between Antibacterial power and chemical structure of HIGHER ALKYL ammonium ions. *J. Bacteriol.* **1945**, *50*, 481–490. [CrossRef] [PubMed]
41. Thilagavathi, G.; Viju, S. Antimicrobials for protective clothing. *Antimicrob. Textiles* **2016**, *6*, 305–317. [CrossRef]
42. Xue, H.; Zhao, Z.; Chen, S.; Du, H.; Chen, R.; Brash, J.L.; Chen, H. Antibacterial coatings based on microgels containing quaternary ammonium ions: Modification with polymeric sugars for improved cytocompatibility. *Colloid Interface Sci. Commun.* **2020**, *37*, 100268. [CrossRef]
43. Pandit, S.; Gaska, K.; Mokkapati, V.R.; Forsberg, S.; Svensson, M.; Kádár, R.; Mijakovic, I. Antibacterial effect of boron nitride flakes with controlled orientation in polymer composites. *RSC Adv.* **2019**, *9*, 33454–33459. [CrossRef]

 *antibiotics* 

Article

Using Lactobacilli to Fight *Escherichia coli* and *Staphylococcus aureus* Biofilms on Urinary Tract Devices

Fábio M. Carvalho, Filipe J. M. Mergulhão and Luciana C. Gomes *

LEPABE—Laboratory for Process Engineering, Environment, Biotechnology and Energy, Faculty of Engineering, University of Porto, Rua Dr. Roberto Frias, 4200-465 Porto, Portugal; up201502963@fe.up.pt (F.M.C.); filipem@fe.up.pt (F.J.M.M.)

* Correspondence: luciana.gomes@fe.up.pt; Tel.: +351-22-041-3603

Abstract: The low efficacy of conventional treatments and the interest in finding natural-based approaches to counteract biofilm development on urinary tract devices have promoted the research on probiotics. This work evaluated the ability of two probiotic strains, *Lactobacillus plantarum* and *Lactobacillus rhamnosus*, in displacing pre-formed biofilms of *Escherichia coli* and *Staphylococcus aureus* from medical-grade silicone. Single-species biofilms of 24 h were placed in contact with each probiotic suspension for 6 h and 24 h, and the reductions in biofilm cell culturability and total biomass were monitored by counting colony-forming units and crystal violet assay, respectively. Both probiotics significantly reduced the culturability of *E. coli* and *S. aureus* biofilms, mainly after 24 h of exposure, with reduction percentages of 70% and 77% for *L. plantarum* and 76% and 63% for *L. rhamnosus*, respectively. Additionally, the amount of *E. coli* biofilm determined by CV staining was maintained approximately constant after 6 h of probiotic contact and significantly reduced up to 67% after 24 h. For *S. aureus*, only *L. rhamnosus* caused a significant effect on biofilm amount after 6 h of treatment. Hence, this study demonstrated the potential of lactobacilli to control the development of pre-established uropathogenic biofilms.

Keywords: biofilm; urinary tract devices; probiotics; *Lactobacillus plantarum*; *Lactobacillus rhamnosus*; antibiofilm activity; displacement

Citation: Carvalho, F.M.; Mergulhão, F.J.M.; Gomes, L.C. Using Lactobacilli to Fight *Escherichia coli* and *Staphylococcus aureus* Biofilms on Urinary Tract Devices. *Antibiotics* **2021**, *10*, 1525. https://doi.org/10.3390/antibiotics10121525

Academic Editors: Andreana Marino and Nicholas Dixon

Received: 30 September 2021
Accepted: 7 December 2021
Published: 14 December 2021

1. Introduction

Urinary tract infections (UTIs) are the most common type of healthcare-associated infections reported by the Centers for Disease Control [1–3], with an estimated annual worldwide incidence of 250 million cases [4]. The high incidence of these infections results in considerable treatment costs, increased length of hospital stays, and high mortality rates, posing a huge financial burden on healthcare systems [5–8]. Device-associated UTIs, caused by the insertion of urological devices (UDs), such as urinary catheters or ureteral stents, contribute to about 97% of UTIs [3,5,9]. Despite the efforts to maintain sterility, the contamination of UDs is almost inevitable since they work as a bridge connecting the nonsterile external environment and the patient's body [10,11]. The most common microorganisms contributing to device-associated UTIs are *Escherichia coli*, *Klebsiella pneumoniae*, *Pseudomonas aeruginosa*, *Staphylococcus aureus*, *Candida* spp., *Enterococcus faecalis*, and *Proteus mirabilis* [12–15].

Device-associated UTIs mostly originate from the formation of microbial pathogenic biofilms on the device's surface [16]. Biofilms are present in about 80% of human microbial infections [12], and once established, they are extremely hard to eliminate [17,18]. Biofilms are defined as a consortium of microorganisms surrounded by a self-synthesized matrix of extracellular polymeric substances [19,20], which protects the embedded bacteria against host defenses and antimicrobial agents [12]. The conditions present in the urinary tract are particularly favorable to microbial adhesion and biofilm development due to the diversity of shear forces prevailing throughout the urinary tract [5,21], the presence of a continuous or intermittent flow of nutrients [12], the absence of defense mechanisms at the UD lumen [12], and the high vulnerability of UDs, classically made of polymeric materials, to bacterial adhesion [22]. In addition, bacteria such as *P. mirabilis* can cause the precipitation of some minerals present in urine, which originates encrustation on UD surfaces and has severe consequences on the bladder and urethral epithelia [9,12].

Some strategies to control UD biofilms include antimicrobial lubricants, bladder instillation or irrigation, antimicrobial agents in collection bags, and the administration of antibiotics [13]. Recently, in a novel approach, the balloon on Foley catheters was transformed into a permeable membrane allowing localized and continuous delivery of antibiotics to the bladder and was shown to eradicate a provoked uropathogenic *E. coli* infection [23]. Moreover, the development of new surfaces that inhibit biofilm formation through antimicrobial agent's release, contact-killing, inhibition of microbial adhesion, and the disruption of biofilm architecture have been suggested to reduce the incidence of device-associated UTIs [24,25]. Although numerous strategies have been investigated, questions regarding biocompatibility, bacterial resistance, long-term efficacy, and cytotoxicity warrants further investigation, not being clear how they will affect clinical outcomes [26].

Recent evidence suggests probiotics as a promising option for fighting biofilms. Probiotics are defined as "live microorganisms which when administered in adequate amounts confer a health benefit on the host" [27]. Lactic acid bacteria (LAB), including *Lactobacillus*, *Bifidobacterium*, *Streptococcus*, *Lactococcus*, and *Leuconostoc*, are the predominant group of bacteria with proven probiotic action [28,29], where *Lactobacillus* assume the greatest relevance [30]. This group of bacteria can grow in different habitats using diverse sources of carbon [31]. From glucose metabolism, LAB are classified as homofermentative, producing exclusively lactic acid, or heterofermentative, producing several other metabolites besides lactic acid, such as ethanol and acetic acid [32,33]. Those substances, together with other secondary metabolites, such as organic acids, exopolysaccharides, biosurfactants, bacteriocins, and enzymes [34], provide a physiologically restrictive environment (e.g., low pH, redox potential, hydrogen sulfide, and peroxide production), making it less suitable for competitors [35–37]. Bacteriocins are a particular class of exometabolites produced by probiotics and are substantially documented to inhibit the growth of competitors [38–40]. Probiotics can also compete for adhesion sites by forming non-pathogenic biofilms that hamper the adhesion and biofilm formation of pathogens [41,42]. Each probiotic strain has multiple and diverse impacts on the host [36]. To date, several studies demonstrated the ability of probiotics to produce antimicrobial metabolites to manage biofilm infections [43–45], and their inhibitory effects on biofilm formation were extensively reviewed [46–48]. Moreover, probiotics were described to suppress quorum-sensing and affect biofilm integrity [46] by repressing the expression of biofilm-associated genes [49].

The objective of this study was to evaluate the effect of two *Lactobacillus* strains frequently used in antimicrobial studies, *Lactobacillus plantarum* (currently *Lactiplantibacillus plantarum*) and *Lactobacillus rhamnosus* (currently *Lacticaseibacillus rhamnosus*), against pre-formed biofilms of bacteria commonly found in biofilms developed in UDs, *E. coli* and *S. aureus*. Some studies showed promising results in displacing adhering uropathogens from catheter materials [16,50,51]. However, to the best of our knowledge, this is the first study that demonstrates the ability of probiotic cells to displace pre-formed biofilms combining the effect of nutritional conditions, temperature, hydrodynamics, and surface material to better predict how probiotics will perform in vivo.

2. Results

To evaluate the capacity of probiotics to disrupt pre-formed pathogenic biofilms, a dynamic biofilm assay was performed where the cell culturability and total biomass were analyzed by colony-forming unit (CFU) count and crystal violet (CV) staining, respectively.

2.1. Biofilm Cell Culturability

The results of *E. coli* and *S. aureus* biofilm culturable cells after 6 h and 24 h of incubation with *L. plantarum* and *L. rhamnosus* suspensions are presented in Figure 1. Both model pathogenic strains confirmed their ability to grow in artificial urine medium, as well as to adhere and form stable biofilms on silicone rubber. The ability of *E. coli* to form robust biofilms on this surface material was previously reported by our research group [52,53]. Moreover, *E. coli* biofilms exhibited higher cellular densities than *S. aureus* (Figure 1), demonstrating its higher propensity to form biofilms on silicone.

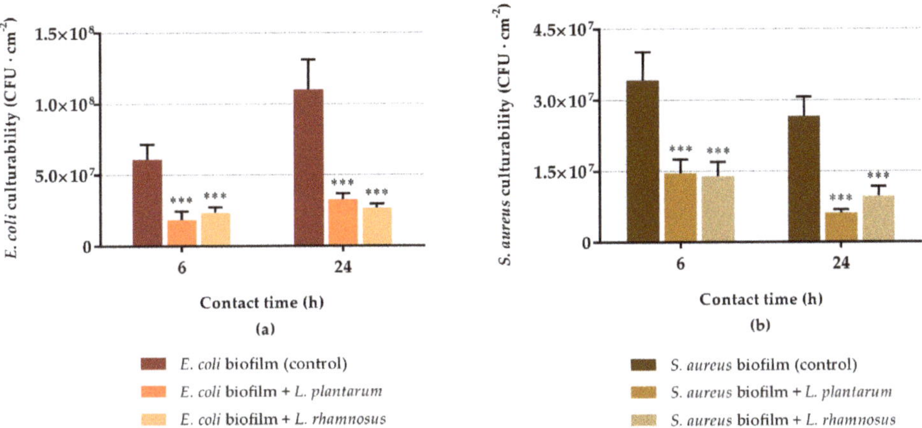

Figure 1. Culturability of 24 h biofilms of *E. coli* (**a**) and *S. aureus* (**b**) after 6 h and 24 h of contact with probiotics (*L. plantarum* or *L. rhamnosus*). The data present the mean and standard deviation (SD) of at least three independent experiments. Statistically significant differences between the control and treatment were considered for *p*-values < 0.05 (*** indicates $p < 0.001$).

Both *Lactobacillus* strains were able to reduce pre-formed biofilms. As regards *E. coli* biofilms (Figure 1a), the number of sessile culturable cells was significantly reduced when exposed to probiotics in comparison to the negative control sample ($p < 0.001$). The highest reductions were obtained after 24 h of exposure, with reductions of 70% for *L. plantarum* and 76% for *L. rhamnosus* ($p < 0.001$). Additionally, reductions of 69% and 61% were obtained after 6 h of biofilm treatment with *L. plantarum* and *L. rhamnosus*, respectively ($p < 0.001$). Regarding *S. aureus* biofilms (Figure 1b), the same tendency was observed. There was a significant reduction in *S. aureus* culturability at both experimental times ($p < 0.001$), being the most significant decrease after 24 h of contact (77% for *L. plantarum* and 63% for *L. rhamnosus*); after 6 h, reductions of 57% and 59% were obtained for *L. plantarum* and *L. rhamnosus*, respectively. For both pathogens, the antimicrobial activity of probiotics increased from 6 h to 24 h of exposure, demonstrating a progressive effect over time.

The culturability of probiotics in the sessile state was simultaneously evaluated (Figure 2). It can be observed that the *E. coli* culturability reduction (Figure 1a) was accompanied by the presence of viable probiotic cells in the biofilms (Figure 2a), except for *L. rhamnosus*, which lost its biofilm culturability after 24 h of interaction with *E. coli*. On the contrary, the reduction in *S. aureus* biofilm culturability (Figure 1b) was not followed by the presence of viable *L. plantarum* cells in biofilms (no colonies were detected at 6 h and 24 h; Figure 2a). *L. rhamnosus* was present in sessile conditions, but its culturability decreased by 85% between 6 h and 24 h of interaction with *S. aureus* (Figure 2b). Additionally, looking at Figures 1 and 2, there is about 1–3 log $CFU{\cdot}cm^{-2}$ difference between the populations of *E. coli* and *S. aureus* and probiotics within the biofilms.

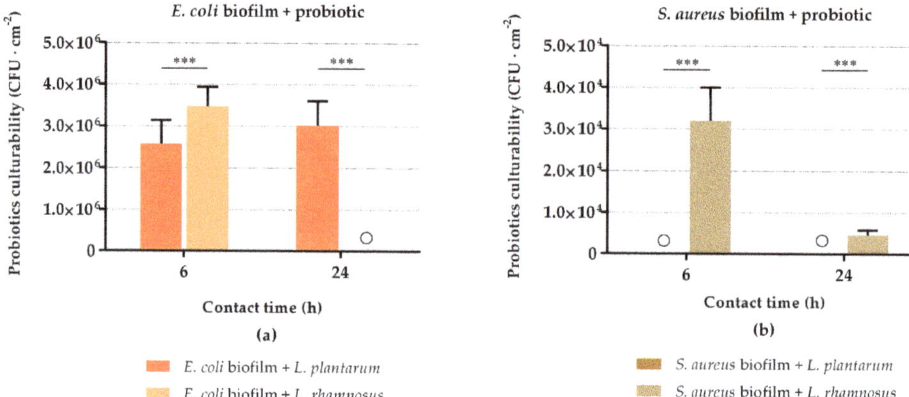

Figure 2. Culturability of attached *L. plantarum* and *L. rhamnosus* cells after 6 h and 24 h of contact with *E. coli* (**a**) and *S. aureus* (**b**) biofilms. The data present the mean and SD of at least three independent experiments. Statistically significant differences between probiotic strains for each time point were considered for p-values < 0.05 (*** indicates $p < 0.001$). "○" indicates that no colonies were detected.

Complementary assays were performed to analyze the presence of probiotic cells in the planktonic fraction in order to explain the decrease in pathogens culturability, even in the absence of *Lactobacillus* cells in the biofilm (data not shown). For example, *L. rhamnosus* cell densities of $2.5 \times 10^5 \pm 4.3 \times 10^3$ and $1.6 \times 10^4 \pm 4.2 \times 10^3$ CFU·mL^{-1} were detected in the suspension after 24 h of contact with *E. coli* and *S. aureus* biofilms, respectively, suggesting that probiotics may act on the pathogenic sessile cells through the release of harmful substances to the extracellular medium. The antimicrobial activity of lactobacilli was also evaluated against *S. aureus* by the disk diffusion method (Supplementary Material). The cell suspensions and cell-free supernatants of both viable and lysed probiotics evidenced clear inhibition zones on the swabbed *S. aureus* on Luria-Bertani agar plates when compared with the negative control (Table S1, Supplementary Material), indicating that these probiotic strains inhibited the growth of pathogens through the production and secretion of antimicrobial substances into their surroundings.

2.2. Biofilm Mass

Figure 3 presents the results of biofilm quantification using CV staining. Concerning *E. coli* biofilms (Figure 3a), both probiotics maintained the biofilm amount at 6 h of contact in comparison with control ($p > 0.05$). However, after 24 h of exposure, *L. plantarum* and *L. rhamnosus* significantly reduced the biofilm mass by 51% ($p = 0.04$) and 67% ($p = 0.011$), respectively, demonstrating their capacity to disrupt the pre-formed biofilms. In the case of *S. aureus* biofilms, the opposite behavior was observed. After 6 h of contact, both probiotics reduced biofilm mass, although only *L. rhamnosus* exhibited statistical difference when compared to control (reduction of 42% for *L. plantarum* ($p = 0.053$) and 20% for *L. rhamnosus* ($p = 0.049$)); at 24 h of contact, the biofilm amount remained nearly constant. Therefore, there was a poor correlation between the CV staining method and cell culturability.

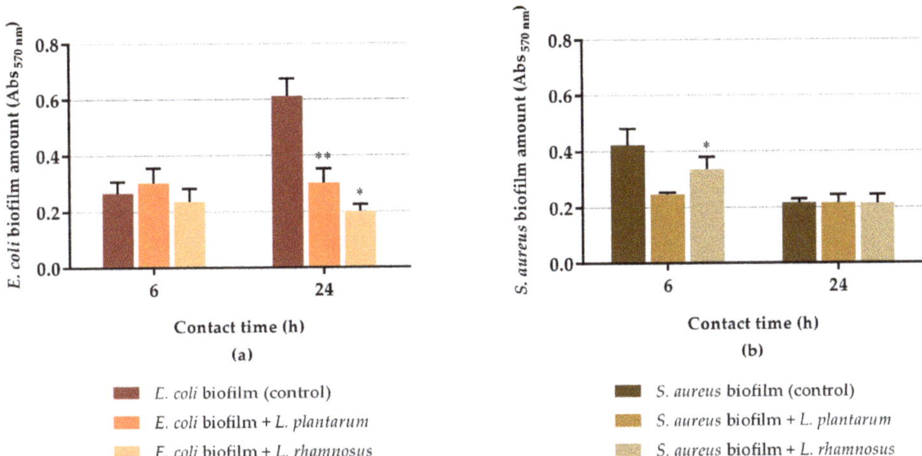

Figure 3. Total amount of *E. coli* (**a**) and *S. aureus* (**b**) biofilms after 6 h and 24 h of contact with probiotics (*L. plantarum* or *L. rhamnosus*). The data present the mean and SD of at least three independent experiments. Statistically significant differences between the control and treatment were considered for *p*-values < 0.05 (* and ** indicate $p < 0.05$ and $p < 0.01$, respectively).

3. Discussion

Device-associated urinary tract infections are a critical problem caused by the high propensity of medical devices to microbial colonization. Previous studies have proposed the use of probiotics as a useful strategy to control pathogenic biofilms, demonstrating that probiotics cells and metabolites can displace adhering uropathogens from urinary devices materials and block bacterial adhesion to uroepithelial cells [47,48]. Probiotics can exert their antibiofilm activity by adopting different strategies: displacement, exclusion, and competition [47]. Recently, our research group evaluated the ability of *L. plantarum* biofilms to prevent *E. coli* adhesion and biofilm formation on silicone rubber, following an exclusion strategy [42]. In the present work, the potential of two probiotic strains (*L. plantarum* and *L. rhamnosus*) to disperse pre-formed biofilms of *E. coli* and *S. aureus* under physiologically relevant conditions was assessed by adopting a displacement strategy [48].

Regarding the antibiofilm activity of probiotics, they significantly inhibited the proliferation of *E. coli* and *S. aureus* biofilms by reducing their cell culturability and biomass amount after 24 h of treatment. Furthermore, both *Lactobacillus* strains caused similar reductions in the culturability of both model pathogens. The activity of probiotics can be related to interfacial cell–cell interactions and to the production and release of antagonizing metabolites that are able to destabilize the biofilm organization, as demonstrated in previous studies [54–56]. Although the incorporation of probiotics within the biofilms suggests that those adhered *Lactobacillus* cells may contribute to pathogen inhibition, the absence of biofilm culturable cells of probiotics at some time points (*L. rhamnosus* after 24 h of contact with *E. coli* and *L. plantarum* after 6 h and 24 h of contact with *S. aureus*) suggests that, beyond the antibiofilm action by integration and contact with the biofilm, lactobacilli may act through the release of antimicrobial substances from the planktonic cells. Moreover, the differences found between the culturability of *E. coli* and *S. aureus* and the culturability of probiotics mean that there was not a direct exchange of pathogenic cells by probiotic cells during treatment. This suggests that pathogen inactivation occurred essentially by the secretion of antimicrobial substances into the surrounding environment. The presence of culturable cells of probiotics in the planktonic fraction and the existence of inhibition zones caused by probiotics cell-free supernatants reinforce this hypothesis.

The poor correlation between the crystal violet (CV) method and cell culturability results can be attributed to the ability of CV to non-specifically bind to some components of the biofilm matrix (such as DNA, exopolysaccharides (EPS), and proteinaceous material) and the peptidoglycan wall of both live and dead cells [57,58].

The antimicrobial substances released by probiotics may penetrate the extracellular matrix of biofilms, interfering with their integrity and cell culturability. Several authors reported the ability of *Lactobacillus* strains to disrupt mature biofilms. Jaffar et al. [58] described that *L. plantarum* significantly disrupted the pre-formed biofilm of *Aggregatibacter actinomycetemcomitans* by approximately 61% on polystyrene after 24 h. Conversely, Song et al. [59] reported that *L. rhamnosus* was capable of disrupting biofilms of *Candida albicans* by 99.9% due to the production of lactic acid and antimicrobial peptides, which presumably disrupted the cytoplasmatic membrane and inactivated cytoplasm molecules. This mechanism was supported by Fayol-Messaoudi et al. [60], who suggested that the bactericidal activity of *Lactobacillus* strains may be due to the synergistic action of lactic acid and secreted bacteriocins. Some studies have shown that bacteriocins produced by *L. plantarum* effectively suppressed the growth and biofilm formation of several microorganisms [39,61,62], including *S. aureus* [63]. McMillan et al. [64] found that *L. rhamnosus* was incorporated into uropathogenic biofilms, including *E. coli*, and caused significant *E. coli* killing supposedly due to the production of bacteriocins or biosurfactant-like substances. In agreement, Cadieux et al. [65] described that *L. rhamnosus* strongly inhibited the growth of uropathogenic *E. coli* by producing bacteriocins, hydrogen peroxide, and lactic acid. Similarly, Otero et al. [45] evidenced an inhibitory effect of *Lactobacillus* strains in *S. aureus* growth after 6 h of co-culture, possibly due to the combined effect of hydrogen peroxide and lactic acid. In addition, some authors suggested that *L. rhamnosus* can produce biosurfactants with antibiofilm activity against *E. coli* and *S. aureus* [43,64]. Previous studies have demonstrated the ability of lactic acid bacteria biosurfactants to inhibit biofilm development and induce its dispersion from surgical implant materials [66–68]. Furthermore, an assorted number of studies attributed the disruption of the architecture of pathogenic biofilms by *L. rhamnosus* to the secretion of molecules that downregulate the genes involved in biofilm development, DNA replication, translation, glycolysis, and gluconeogenesis [55,56]. Song et al. [69] reported that *L. rhamnosus* significantly disrupted the architecture of *E. coli* biofilms by 82% by decreasing the transcriptional activity of transcriptional activators (*luxS*, *lsrK*, and *lsrR*) of the quorum-sensing in *E. coli*. In addition, Ahn et al. [70] and Kim et al. [71] described that lipoteichoic acid produced by *L. plantarum* disrupted pre-formed biofilms of single and multispecies pathogens by interfering with EPS production.

In this work, the lack of glucose in the growth medium hindered probiotics fermentation since high glucose concentrations enhance medium acidification [72,73]. This was confirmed by comparing the initial and final pH values of artificial urine medium (AUM), where no differences were found. Additionally, Todorov et al. [74] studied the effect of the pH on the production of bacteriocins of *L. plantarum* and found that the inhibitory activity of bacteriocins was detected at pH values between 5.0 and 6.5. Moreover, the presence of yeast extract, potassium dihydrogen phosphate, and dipotassium hydrogen phosphate in culture medium as nitrogen and phosphorus sources, respectively, enhanced the production of bacteriocins by *L. plantarum* [74]. In the present work, since the initial pH of AUM is 6.5, and yeast extract, potassium dihydrogen phosphate, and dipotassium hydrogen phosphate are present in AUM [75], this may have contributed to the production of bacteriocins by this probiotic strain. Regarding biosurfactants, despite the bactericidal potential of biosurfactants, their action is more frequently associated with their capacity to affect microbial adhesion by interfering with surface tension and hydrophobicity [76,77], so biosurfactants are unlikely to be effective on mature biofilms. Thus, the disruptive activity of probiotics is most likely explained by the production and secretion of antimicrobial exometabolites, such as hydrogen peroxide, bacteriocins, and biosurfactants, rather than by a possible pH change of the culture media.

Overall, these results suggest that the ability of probiotics to displace pre-formed uropathogenic biofilms can be attributed to the production of exometabolites that inactivate sessile cells and destabilize the biofilm structure.

4. Materials and Methods

4.1. Preparation of Silicone Surfaces

Biofilm formation experiments were performed on silicone coupons (1 × 1 cm; Neves & Neves, Lda, Maia, Portugal) with the intention of mimicking the most common material of urinary catheters [78]. The surfaces were prepared as previously described by Carvalho et al. [42]. Briefly, the coupons were washed with 70% (*v*/*v*) ethanol (VWR, Radnor, PA, USA), air-dried, and then sterilized through ultra-violet (UV) radiation for 30 min. The sterile coupons were fixed to the bottom of 12-well polystyrene plates (VWR, USA) using double-sided adhesive tape, which was sterilized beforehand for 30 min through UV radiation.

4.2. Bacterial Strains and Culture Conditions

The probiotic strains used in this study were *Lactobacillus plantarum* and *Lactobacillus rhamnosus* (1×10^{11} CFU·g^{-1}; Biomodics ApS, Rødovre, Denmark). These bacteria were preserved at −80 °C in De Man, Rogosa and Sharpe (MRS) broth (Merck KGaA, Madrid, Spain) with 30% (*v*/*v*) glycerol, streaked on MRS agar (Scharlab, S.L., Barcelona, Spain) plates, and incubated for 48 h at 37 °C. *Lactobacillus* inocula were prepared by collecting bacterial colonies from the MRS agar plate into 250 mL of MRS broth and incubating overnight at 37 °C in an orbital shaker at 120 rpm (Agitorb 200, Aralab, Rio de Mouro, Portugal). MRS is the culture media routinely used for lactobacilli growth [79].

Escherichia coli CECT 434 and *Staphylococcus aureus* CECT 976 were chosen as model microorganisms of biofilm-based urinary tract infections. These bacterial strains were preserved at −80 °C in Luria-Bertani (LB) broth (Thermo Fisher Scientific, Waltham, MA, USA) containing 30% (*v*/*v*) glycerol, streaked on LB agar plates, and incubated for 24 h at 37 °C. The starting cultures were prepared by collecting single colonies from LB agar plates to 250 mL of artificial urine medium (AUM) [75] and incubating overnight at 37 °C and 120 rpm. AUM was used to simulate the nutrient composition of human urine [75].

4.3. Influence of Probiotics on Pre-Formed Biofilms

The antibiofilm assays followed a displacement strategy [47], which consisted of the formation of *E. coli* and *S. aureus* biofilms, after which the biofilms were inoculated with the *Lactobacillus* strains separately in order to evaluate their ability to disperse pre-formed biofilms. The bacterial cultures grown overnight were harvested by centrifugation at 3202× *g* for 10 min at 25 °C (Eppendorf Centrifuge 5810R, Hamburg, Germany), and the final cell concentration was adjusted in fresh AUM to an optical density at 610 nm of 0.15 for *E. coli*, 0.20 for *S. aureus*, and 0.70 for both *Lactobacillus* strains, equivalent to approximately 10^8 CFU·mL^{-1}, the recommended bacterial density to be used in urological experiments [5]. This was confirmed by colony-forming unit (CFU) counts. The uropathogenic biofilms were formed on silicone coupons placed inside 12-well plates where each well was filled with 3 mL of the respective pathogenic suspension. The plates were incubated for 24 h at 37 °C under shaking conditions in order to generate shear stresses similar to those found inside urinary catheters [42,80]. Afterwards, cell suspensions were removed, and each well was loaded with 3 mL of the respective probiotic suspension for periods of contact of 6 h and 24 h under the same growth conditions. A negative control was prepared by adding sterile AUM to pathogenic biofilms. At the end of each experimental period, biofilms were analyzed as previously described [42]. Briefly, cell suspensions were removed from the wells, the non-adherent cells were washed with a sodium chloride solution (8.5 g·L^{-1} NaCl), and the biofilm amount and culturability were determined by crystal violet (CV) staining and CFU counts, respectively. All experiments included at least three independent biological replicates with two technical replicates each.

4.3.1. Bacterial Enumeration

The number of biofilm culturable cells per cm^2 of silicone was determined as indicated by Carvalho et al. [42]. Briefly, the coupons were transferred to Falcon tubes with 2 mL of saline solution and the biofilm cells were detached from the coupons by vortexing (ZX4, Velp Scientifica, Usmate Velate, Italy) for 2 min at full power. Then, serial dilutions of the obtained biofilm cell suspensions were performed in saline solution, plated on LB agar (selective media for *E. coli* and *S. aureus*) and MRS agar (selective media for *L. plantarum* and *L. rhamnosus*), and incubated at 37 °C for 24 h and 48 h, respectively.

The percentages of CFU reduction in pathogens were estimated as follows:

$$\text{Reduction } (\%) = \frac{\text{CFU}_{\text{control}} - \text{CFU}_{\text{biofilm}}}{\text{CFU}_{\text{control}}} \times 100 \quad (1)$$

where $\text{CFU}_{\text{control}}$ corresponds to the number of culturable cells of pathogens in the negative control samples (biofilms not exposed to probiotic cell suspensions), and $\text{CFU}_{\text{biofilm}}$ is the number of culturable cells in biofilms treated with probiotics.

4.3.2. Biofilm Amount Determination

The total mass of biofilms was quantified using the CV staining method [42]. Briefly, after washing the non-adherent cells, silicone coupons were transferred to 24-well polystyrene plates (Thermo Fisher Scientific, USA), and biofilms were fixed with 1 mL of 100% ethanol (VWR, USA) for 15 min. Then, the wells were air-dried, and biofilms were stained with 1 mL of 1% (*v*/*v*) CV (Merck, Germany) solution for 5 min. The dye bounded to the biofilm was solubilized by adding 1 mL of 33% (*v*/*v*) acetic acid (VWR, USA) solution. Finally, 200 μL of each well was transferred to a 96-well polystyrene plate (VWR, USA), and the biofilm mass was determined through absorbance measurement at 570 nm ($\text{Abs}_{570\text{ nm}}$) in a microtiter plate reader (SpectroStar Nano, Biogen Cientifica S. L., Madrid, Spain). When the absorbance values exceeded 1, samples were diluted in 33% (*v*/*v*) acetic acid. The biofilm amount was expressed as $\text{Abs}_{570\text{ nm}}$ values.

4.4. Statistical Analysis

Statistical analysis was performed using the IBM SPSS Statistics version 26 for Windows (IBM SPSS, Inc., Chicago, IL, USA). Descriptive statistics were used to calculate the mean and standard deviation (SD) for the number of culturable cells and biofilm mass. The homogeneity of variances and normality of data were verified for all response variables tested using the Kolmogorov–Smirnov and Shapiro–Wilk tests. Since the response variables were not normally distributed, a nonparametric analysis using the Kruskal–Wallis test was performed to assess whether there were statistically significant differences among groups, and the differences between those groups were determined by the Mann–Whitney test. Statistically significant differences were considered for *p*-values < 0.05, corresponding to a confidence level of 95% (*, **, and *** indicate $p < 0.05$, $p < 0.01$ and $p < 0.001$, respectively). All reported data are presented as mean ± SD from at least three experiments with duplicates.

5. Conclusions

L. plantarum and *L. rhamnosus* were able to displace pre-established biofilms of *E. coli* and *S. aureus* at similar levels. The antibiofilm activity of these probiotic strains may be primarily linked to the release of self-produced substances, which reduced the number of culturable pathogens in biofilms. Additionally, the integration of probiotic cells into the biofilm may have contributed to the destabilization of the biofilm organization.

This proof-of-principle study supports the potential of *Lactobacillus* strains to be used as biocontrol agents against pathogenic biofilms developed on urinary tract devices. It will pave the way for more experiments on the topic in an effort to elucidate the mechanisms underlying the lactobacilli activity.

Supplementary Materials: The following are available online at https://www.mdpi.com/article/10.3390/antibiotics10121525/s1. Detection of Antimicrobial Compounds by Growth Inhibition on Agar Plate; Table S1: The inhibitory activity of viable and lysed *L. plantarum* and *L. rhamnosus* suspensions and cell-free supernatants on the growth of *S. aureus*.

Author Contributions: Conceptualization, F.J.M.M. and L.C.G.; methodology, F.M.C. and L.C.G.; software, F.M.C. and L.C.G.; investigation, F.M.C. and L.C.G.; resources, F.J.M.M. and L.C.G.; data curation, F.M.C. and L.C.G.; writing—original draft preparation, F.M.C.; writing—review and editing, F.M.C., F.J.M.M. and L.C.G.; supervision, F.J.M.M. and L.C.G. All authors have read and agreed to the published version of the manuscript.

Funding: This research was funded by: Base Funding—UIDB/00511/2020 of the Laboratory for Process Engineering, Environment, Biotechnology and Energy—LEPABE—funded by national funds through the FCT/MCTES (PIDDAC); Project PTDC/CTMCOM/4844/2020 funded by the Portuguese Foundation for Science and Technology (FCT). L.C.G. thanks the FCT for financial support of her work contract through the Scientific Employment Stimulus-Individual Call-[CEECIND/01700/2017].

Institutional Review Board Statement: Not applicable.

Informed Consent Statement: Not applicable.

Data Availability Statement: The data presented in this study are available on request from the corresponding author. The data are not publicly available yet as some data sets are being used for additional publications.

Acknowledgments: We thank Martin Alm and Peter Thomsen (Biomodics ApS, Denmark) for providing free of charge the two probiotic strains (*L. plantarum* and *L. rhamnosus*). Mette Burmølle from the Department of Biology, University of Copenhagen, is also acknowledged for her support throughout this work.

Conflicts of Interest: The authors declare no conflict of interest.

References

1. Centers for Disease Control and Prevention Catheter-associated Urinary Tract Infections (CAUTI) | HAI | CDC. Available online: https://www.cdc.gov/hai/ca_uti/uti.html (accessed on 20 September 2021).
2. Siddiq, D.M.; Darouiche, R.O. New strategies to prevent catheter-associated urinary tract infections. *Nat. Rev. Urol.* **2012**, *9*, 305–314. [CrossRef]
3. Maharjan, G.; Khadka, P.; Siddhi Shilpakar, G.; Chapagain, G.; Dhungana, G.R. Catheter-Associated Urinary Tract Infection and Obstinate Biofilm Producers. *Can. J. Infect. Dis. Med. Microbiol.* **2018**, *2018*, 7624857. [CrossRef]
4. World Health Organization Report on the Burden of Endemic Health Care-Associated Infection Worldwide. 2011. Available online: https://apps.who.int/iris/handle/10665/80135 (accessed on 20 September 2021).
5. Ramstedt, M.; Ribeiro, I.A.C.; Bujdakova, H.; Mergulhão, F.J.M.; Jordao, L.; Thomsen, P.; Alm, M.; Burmølle, M.; Vladkova, T.; Can, F.; et al. Evaluating efficacy of antimicrobial and antifouling materials for urinary tract medical devices: Challenges and recommendations. *Macromol. Biosci.* **2019**, *19*, e1800384. [CrossRef]
6. Percival, S.L.; Suleman, L.; Vuotto, C.; Donelli, G. Healthcare-associated infections, medical devices and biofilms: Risk, tolerance and control. *J. Med. Microbiol.* **2015**, *64*, 323–334. [CrossRef] [PubMed]
7. Khatoon, Z.; McTiernan, C.D.; Suuronen, E.J.; Mah, T.F.; Alarcon, E.I. Bacterial biofilm formation on implantable devices and approaches to its treatment and prevention. *Heliyon* **2018**, *4*, e01067. [CrossRef]
8. Lima, M.; Teixeira-Santos, R.; Gomes, L.C.; Faria, S.I.; Valcarcel, J.; Vázquez, J.A.; Cerqueira, M.A.; Pastrana, L.; Bourbon, A.I.; Mergulhão, F.J. Development of Chitosan-Based Surfaces to Prevent Single- and Dual-Species Biofilms of Staphylococcus aureus and Pseudomonas aeruginosa. *Molecules* **2021**, *26*, 4378. [CrossRef] [PubMed]
9. Mandakhalikar, K.D.; Chua, R.R.; Tambyah, P.A. New Technologies for Prevention of Catheter Associated Urinary Tract Infection. *Curr. Treat. Options Infect. Dis.* **2016**, *8*, 24–41. [CrossRef]
10. Tunney, M.M.; Gorman, S.P.; Patrick, S. Infection associated with medical devices. *Int. J. Gen. Syst.* **2002**, *31*, 195–205. [CrossRef]
11. Vertes, A.; Hitchins, V.; Phillips, K.S. Analytical challenges of microbial biofilms on medical devices. *Anal. Chem.* **2012**, *84*, 3858–3866. [CrossRef] [PubMed]
12. Azevedo, A.S.; Almeida, C.; Melo, L.F.; Azevedo, N.F. Impact of polymicrobial biofilms in catheter-associated urinary tract infections. *Crit. Rev. Microbiol.* **2017**, *43*, 423–439. [CrossRef]
13. Donlan, R.M. Biofilms and device-associated infections. *Emerg. Infect. Dis.* **2001**, *7*, 277–281. [CrossRef]
14. Seif Eldein, S.S.; El-Temawy, A.-E.-K.A.; Ahmed, E.H. Biofilm Formation by *E. coli* Causing Catheter Associated Urinary Tract Infection (CAUTI) in Assiut University Hospital. *Egypt. J. Med. Microbiol.* **2013**, *22*, 101–110. [CrossRef]

15. Niveditha, S.; Pramodhini, S.; Umadevi, S.; Kumar, S.; Stephen, S. The isolation and the biofilm formation of uropathogens in the patients with catheter associated urinary tract infections (UTIs). *J. Clin. Diagn. Res.* **2012**, *6*, 1478–1482. [CrossRef] [PubMed]
16. Chen, Q.; Zhu, Z.; Wang, J.; Lopez, A.I.; Li, S.; Kumar, A.; Yu, F.; Chen, H.; Cai, C.; Zhang, L. Probiotic *E. coli* Nissle 1917 biofilms on silicone substrates for bacterial interference against pathogen colonization. *Acta Biomater.* **2017**, *50*, 353–360. [CrossRef] [PubMed]
17. Vlamakis, H.; Kolter, R. Biofilms. *Cold Spring Harb. Perspect. Biol.* **2010**, *2*, a000398.
18. Donlan, R.M. Biofilms: Microbial life on surfaces. *Emerg. Infect. Dis.* **2002**, *8*, 881–890. [CrossRef]
19. Flemming, H.C.; Wingender, J. The biofilm matrix. *Nat. Rev. Microbiol.* **2010**, *8*, 623–633. [CrossRef]
20. Rabin, N.; Zheng, Y.; Opoku-Temeng, C.; Du, Y.; Bonsu, E.; Sintim, H.O. Biofilm formation mechanisms and targets for developing antibiofilm agents. *Future Med. Chem.* **2015**, *7*, 493–512. [CrossRef]
21. Schembri, M.A.; Klemm, P. Biofilm formation in a hydrodynamic environment by novel FimH variants and ramifications for virulence. *Infect. Immun.* **2001**, *69*, 1322–1328. [CrossRef]
22. Lawrence, E.L.; Turner, I.G. Materials for urinary catheters: A review of their history and development in the UK. *Med. Eng. Phys.* **2005**, *27*, 443–453. [CrossRef]
23. Stærk, K.; Grønnemose, R.B.; Palarasah, Y.; Kolmos, H.J.; Lund, L.; Alm, M.; Thomsen, P.; Andersen, T.E. A Novel Device-Integrated Drug Delivery System for Local Inhibition of Urinary Tract Infection. *Front. Microbiol.* **2021**, *12*, 1618. [CrossRef]
24. Chen, M.; Yu, Q.; Sun, H. Novel strategies for the prevention and treatment of biofilm related infections. *Int. J. Mol. Sci.* **2013**, *14*, 18488–18501. [CrossRef]
25. Zhu, Z.; Wang, Z.; Li, S.; Yuan, X. Antimicrobial strategies for urinary catheters. *J. Biomed. Mater. Res.—Part A* **2019**, *107*, 445–467. [CrossRef] [PubMed]
26. Singha, P.; Locklin, J.; Handa, H. A review of the recent advances in antimicrobial coatings for urinary catheters. *Acta Biomater.* **2017**, *50*, 20–40. [CrossRef] [PubMed]
27. AGN FAO; Nutrition and Consumer Protection Div; WHO, Geneva. *FAO Probiotics in Food: Health and Nutritional Properties and Guidelines for Evaluation*; FAO: Rome, Italy, 2006.
28. Fioramonti, J.; Theodorou, V.; Bueno, L. Probiotics: What are they? What are their effects on gut physiology? *Best Pract. Res. Clin. Gastroenterol.* **2003**, *17*, 711–724. [CrossRef]
29. Gogineni, V.K.; Morrow, L.E. Probiotics: Mechanisms of action and clinical applications. *J. Probiotics Heal.* **2013**, *1*, 101. [CrossRef]
30. Aoudia, N.; Rieu, A.; Briandet, R.; Deschamps, J.; Chluba, J.; Jego, G.; Garrido, C.; Guzzo, J. Biofilms of *Lactobacillus plantarum* and *Lactobacillus fermentum*: Effect on stress responses, antagonistic effects on pathogen growth and immunomodulatory properties. *Food Microbiol.* **2016**, *53*, 51–59. [CrossRef] [PubMed]
31. Muñoz, M.; Mosquera, A.; Alméciga-Díaz, C.J.; Melendez, A.P.; Sánchez, O.F. Fructooligosaccharides metabolism and effect on bacteriocin production in *Lactobacillus* strains isolated from ensiled corn and molasses. *Anaerobe* **2012**, *18*, 321–330. [CrossRef]
32. de Melo Pereira, G.V.; de Oliveira Coelho, B.; Magalhães Júnior, A.I.; Thomaz-Soccol, V.; Soccol, C.R. How to select a probiotic? A review and update of methods and criteria. *Biotechnol. Adv.* **2018**, *36*, 2060–2076. [CrossRef]
33. Carr, F.J.; Chill, D.; Maida, N. The lactic acid bacteria: A literature survey. *Crit. Rev. Microbiol.* **2002**, *28*, 281–370. [CrossRef]
34. Prabhurajeshwar, C.; Chandrakanth, R.K. Probiotic potential of lactobacilli with antagonistic activity against pathogenic strains: An in vitro validation for the production of inhibitory substances. *Biomed. J.* **2017**, *40*, 270–283. [CrossRef]
35. Bermudez-Brito, M.; Plaza-Díaz, J.; Muñoz-Quezada, S.; Gómez-Llorente, C.; Gil, A. Probiotic mechanisms of action. *Ann. Nutr. Metab.* **2012**, *61*, 160–174. [CrossRef]
36. Ng, S.C.; Hart, A.L.; Kamm, M.A.; Stagg, A.J.; Knight, S.C. Mechanisms of action of probiotics: Recent advances. *Inflamm. Bowel Dis.* **2009**, *15*, 300–310. [CrossRef]
37. Khalighi, A.; Behdani, R.; Kouhestani, S. Probiotics: A comprehensive review of their classification, mode of action and role in human nutrition. *Probiotics Prebiotics Hum. Nutr. Health.* **2016**, *10*, 63646.
38. Ray Mohapatra, A.; Jeevaratnam, K. Inhibiting bacterial colonization on catheters: Antibacterial and antibiofilm activities of bacteriocins from *Lactobacillus plantarum* SJ33. *J. Glob. Antimicrob. Resist.* **2019**, *19*, 85–92. [CrossRef]
39. Hasslöf, P.; Hedberg, M.; Twetman, S.; Stecksén-Blicks, C. Growth inhibition of oral *mutans* streptococci and *candida* by commercial probiotic lactobacilli—An *in vitro* study. *BMC Oral Health* **2010**, *10*, 18. [CrossRef]
40. Vahedi Shahandashti, R.; Kasra Kermanshahi, R.; Ghadam, P. The inhibitory effect of bacteriocin produced by *Lactobacillus acidophilus* ATCC 4356 and *Lactobacillus plantarum* ATCC 8014 on planktonic cells and biofilms of *Serratia marcescens*. *Turkish J. Med. Sci.* **2016**, *46*, 1188–1196. [CrossRef]
41. Jalilsood, T.; Baradaran, A.; Song, A.A.L.; Foo, H.L.; Mustafa, S.; Saad, W.Z.; Yusoff, K.; Rahim, R.A. Inhibition of pathogenic and spoilage bacteria by a novel biofilm-forming *Lactobacillus* isolate: A potential host for the expression of heterologous proteins. *Microb. Cell Fact.* **2015**, *14*, 96. [CrossRef]
42. Carvalho, F.M.; Teixeira-Santos, R.; Mergulhão, F.J.M.; Gomes, L.C. Effect of *Lactobacillus plantarum* Biofilms on the Adhesion of *Escherichia coli* to Urinary Tract Devices. *Antibiotics* **2021**, *10*, 966. [CrossRef]
43. Sambanthamoorthy, K.; Feng, X.; Patel, R.; Patel, S.; Paranavitana, C. Antimicrobial and antibiofilm potential of biosurfactants isolated from lactobacilli against multi-drug-resistant pathogens. *BMC Microbiol.* **2014**, *14*, 197. [CrossRef]
44. Kaur, S.; Sharma, P.; Kalia, N.; Singh, J.; Kaur, S. Anti-biofilm properties of the fecal probiotic lactobacilli against *Vibrio* spp. *Front. Cell. Infect. Microbiol.* **2018**, *8*, 120. [CrossRef]

45. Otero, M.C.; Nader-Macías, M.E. Inhibition of *Staphylococcus aureus* by H2O2-producing *Lactobacillus gasseri* isolated from the vaginal tract of cattle. *Anim. Reprod. Sci.* **2006**, *96*, 35–46. [CrossRef]
46. Barzegari, A.; Kheyrolahzadeh, K.; Mahdi, S.; Khatibi, H.; Sharifi, S.; Memar, M.Y.; Vahed, S.Z. The battle of probiotics and their derivatives against biofilms. *Infect. Drug Resist.* **2020**, *13*, 659–672. [CrossRef]
47. Carvalho, F.M.; Teixeira-Santos, R.; Mergulhão, F.J.M.; Gomes, L.C. The use of probiotics to fight biofilms in medical devices: A systematic review and meta-analysis. *Microorganisms* **2021**, *9*, 27. [CrossRef]
48. Carvalho, F.M.; Teixeira-Santos, R.; Mergulhão, F.J.M.; Gomes, L.C. Targeting biofilms in medical devices using probiotic cells: A systematic review. *AIMS Mater. Sci.* **2021**, *8*, 501–523. [CrossRef]
49. Jeong, D.; Kim, D.H.; Song, K.Y.; Seo, K.H. Antimicrobial and anti-biofilm activities of *Lactobacillus kefiranofaciens* DD2 against oral pathogens. *J. Oral Microbiol.* **2018**, *10*, 1472985. [CrossRef]
50. Cadieux, P.; Watterson, J.D.; Denstedt, J.; Harbottle, R.R.; Puskas, J.; Howard, J.; Gan, B.S.; Reid, G. Potential application of polyisobutylene-polystyrene and a *Lactobacillus* protein to reduce the risk of device-associated urinary tract infections. *Colloids Surf B Biointerfaces* **2003**, *28*, 95–105. [CrossRef]
51. Reid, G.; Tieszer, C. Use of lactobacilli to reduce the adhesion of *Staphylococcus aureus* to catheters. *Int. Biodeterior. Biodegrad.* **1994**, *34*, 73–83. [CrossRef]
52. Gomes, L.C.; Silva, L.N.; Simões, M.; Melo, L.F.; Mergulhão, F.J. *Escherichia coli* adhesion, biofilm development and antibiotic susceptibility on biomedical materials. *J. Biomed. Mater. Res. Part A* **2015**, *103*, 1414–1423. [CrossRef]
53. Azevedo, A.S.; Almeida, C.; Gomes, L.C.; Ferreira, C.; Mergulhão, F.J.; Melo, L.F.; Azevedo, N.F. An *in vitro* model of catheter-associated urinary tract infections to investigate the role of uncommon bacteria on the *Escherichia coli* microbial consortium. *Biochem. Eng. J.* **2017**, *118*, 64–69. [CrossRef]
54. Tan, Y.; Leonhard, M.; Moser, D.; Ma, S.; Schneider-Stickler, B. Inhibitory effect of probiotic lactobacilli supernatants on single and mixed non-albicans *Candida* species biofilm. *Arch. Oral Biol.* **2018**, *85*, 40–45. [CrossRef]
55. Matsubara, V.H.; Wang, Y.; Bandara, H.M.H.N.; Mayer, M.P.A.; Samaranayake, L.P. Probiotic lactobacilli inhibit early stages of *Candida albicans* biofilm development by reducing their growth, cell adhesion, and filamentation. *Appl. Microbiol. Biotechnol.* **2016**, *100*, 6415–6426. [CrossRef]
56. Rossoni, R.D.; de Barros, P.P.; de Alvarenga, J.A.; de Camargo Ribeiro, F.; dos Santos Velloso, M.; Fuchs, B.B.; Mylonakis, E.; Jorge, A.O.C.; Junqueira, J.C. Antifungal activity of clinical *Lactobacillus* strains against *Candida albicans* biofilms: Identification of potential probiotic candidates to prevent oral candidiasis. *Biofouling* **2018**, *34*, 212–225. [CrossRef]
57. Fernández Ramírez, M.D.; Smid, E.J.; Abee, T.; Nierop Groot, M.N. Characterisation of biofilms formed by *Lactobacillus plantarum* WCFS1 and food spoilage isolates. *Int. J. Food Microbiol.* **2015**, *207*, 23–29. [CrossRef]
58. Jaffar, N.; Ishikawa, Y.; Mizuno, K.; Okinaga, T.; Maeda, T. Mature biofilm degradation by potential probiotics: *Aggregatibacter actinomycetemcomitans* versus *Lactobacillus* spp. *PLoS ONE* **2016**, *11*, e0159466. [CrossRef]
59. Song, Y.G.; Lee, S.H. Inhibitory effects of *Lactobacillus rhamnosus* and *Lactobacillus casei* on *Candida* biofilm of denture surface. *Arch. Oral Biol.* **2017**, *76*, 1–6. [CrossRef]
60. Fayol-Messaoudi, D.; Berger, C.N.; Coconnier-Polter, M.H.; Liévin-Le Moal, V.; Servin, A.L. pH-, lactic acid-, and non-lactic acid-dependent activities of probiotic lactobacilli against *Salmonella enterica* serovar *typhimurium*. *Appl. Environ. Microbiol.* **2005**, *71*, 6008–6013. [CrossRef]
61. Lin, X.; Chen, X.; Tu, Y.; Wang, S.; Chen, H. Effect of probiotic lactobacilli on the growth of *Streptococcus mutans* and multispecies biofilms isolated from children with active caries. *Med. Sci. Monit.* **2017**, *23*, 4175–4181. [CrossRef]
62. Maldonado-Barragán, A.; Caballero-Guerrero, B.; Lucena-Padrós, H.; Ruiz-Barba, J.L. Induction of bacteriocin production by coculture is widespread among plantaricin-producing *Lactobacillus plantarum* strains with different regulatory operons. *Food Microbiol.* **2013**, *33*, 40–47. [CrossRef]
63. Klaenhammer, T.R. Bacteriocins of lactic acid bacteria. *Biochimie* **1988**, *70*, 337–349. [CrossRef]
64. McMillan, A.; Dell, M.; Zellar, M.P.; Cribby, S.; Martz, S.; Hong, E.; Fu, J.; Abbas, A.; Dang, T.; Miller, W.; et al. Disruption of urogenital biofilms by lactobacilli. *Colloids Surf. B Biointerfaces* **2011**, *86*, 58–64. [CrossRef] [PubMed]
65. Cadieux, P.A.; Burton, J.P.; Devillard, E.; Reid, G. *Lactobacillus* by-products inhibit the growth and virulence of uropathogenic *Escherichia coli*. *J. Physiol. Pharmacol.* **2009**, *60*, 13–18. [PubMed]
66. Morais, I.M.C.; Cordeiro, A.L.; Teixeira, G.S.; Domingues, V.S.; Nardi, R.M.D.; Monteiro, A.S.; Alves, R.J.; Siqueira, E.P.; Santos, V.L. Biological and physicochemical properties of biosurfactants produced by *Lactobacillus jensenii* P6A and *Lactobacillus gasseri* P65. *Microb. Cell Fact.* **2017**, *16*, 155. [CrossRef] [PubMed]
67. Ceresa, C.; Tessarolo, F.; Caola, I.; Nollo, G.; Cavallo, M.; Rinaldi, M.; Fracchia, L. Inhibition of *Candida albicans* adhesion on medical-grade silicone by a *Lactobacillus*-derived biosurfactant. *J. Appl. Microbiol.* **2015**, *118*, 1116–1125. [CrossRef]
68. Sharma, D.; Saharan, B.S. Functional characterization of biomedical potential of biosurfactant produced by *Lactobacillus helveticus*. *Biotechnol. Reports* **2016**, *11*, 27–35. [CrossRef]
69. Song, H.; Zhang, J.; Qu, J.; Liu, J.; Yin, P.; Zhang, G.; Shang, D. *Lactobacillus rhamnosus* GG microcapsules inhibit *Escherichia coli* biofilm formation in coculture. *Biotechnol. Lett.* **2019**, *41*, 1007–1014. [CrossRef]
70. Ahn, K.B.; Baik, J.E.; Park, O.J.; Yun, C.H.; Han, S.H. *Lactobacillus plantarum* lipoteichoic acid inhibits biofilm formation of *Streptococcus mutans*. *PLoS ONE* **2018**, *13*, e0192694. [CrossRef]

71. Kim, A.R.; Ahn, K.B.; Yun, C.H.; Park, O.J.; Perinpanayagam, H.; Yoo, Y.J.; Kum, K.Y.; Han, S.H. *Lactobacillus plantarum* lipoteichoic acid inhibits oral multispecies biofilm. *J. Endod.* **2019**, *45*, 310–315. [CrossRef]
72. Teanpaisan, R.; Piwat, S.; Dahlén, G. Inhibitory effect of oral *Lactobacillus* against oral pathogens. *Lett. Appl. Microbiol.* **2011**, *53*, 452–459. [CrossRef]
73. Alexandre, Y.; Le Berre, R.; Barbier, G.; Le Blay, G. Screening of *Lactobacillus* spp. for the prevention of *Pseudomonas aeruginosa* pulmonary infections. *BMC Microbiol.* **2014**, *14*, 107. [CrossRef]
74. Todorov, S.; Gotcheva, B.; Dousset, X.; Onno, B.; Ivanova, I. Influence of growth medium on bacteriocin production in *Lactobacillus plantarum* ST31. *Biotechnol. Biotechnol. Equip.* **2000**, *14*, 50–55. [CrossRef]
75. Brooks, T.; Keevil, C.W. A simple artificial urine for the growth of urinary pathogens. *Lett. Appl. Microbiol.* **1997**, *24*, 203–206. [CrossRef]
76. Rodrigues, L.; Banat, I.M.; Teixeira, J.; Oliveira, R. Biosurfactants: Potential applications in medicine. *J. Antimicrob. Chemother.* **2006**, *57*, 609–618. [CrossRef] [PubMed]
77. Fracchia, L.; Cavallo, M.; Giovanna, M.; Banat, I.M. Biosurfactants and Bioemulsifiers Biomedical and Related Applications—Present Status and Future Potentials. In *Biomedical Science, Engineering and Technology*; Ghista, D.N., Ed.; InTech Publisher: London, UK, 2012; pp. 325–370, ISBN 978-953-307-471-9.
78. Cerqueira, L.; Oliveira, J.A.; Nicolau, A.; Azevedo, N.F.; Vieira, M.J. Biofilm formation with mixed cultures of *Pseudomonas aeruginosa/Escherichia coli* on silicone using artificial urine to mimic urinary catheters. *Biofouling* **2013**, *29*, 829–840. [CrossRef]
79. Leroy, F.; De Vuyst, L. Growth of the Bacteriocin-Producing *Lactobacillus sakei* Strain CTC 494 in MRS Broth is Strongly Reduced Due to Nutrient Exhaustion: A Nutrient Depletion Model for the Growth of Lactic Acid Bacteria. *Appl. Environ. Microbiol.* **2001**, *67*, 4407–4413. [CrossRef] [PubMed]
80. Gomes, M.; Gomes, L.C.; Teixeira-Santos, R.; Pereira, M.F.R.; Soares, O.S.G.P.; Mergulhão, F.J. Optimizing CNT Loading in Antimicrobial Composites for Urinary Tract Application. *Appl. Sci.* **2021**, *11*, 4038. [CrossRef]

 *antibiotics*

Review

Antimicrobial and Antibiofilm Coating of Dental Implants—Past and New Perspectives

Guilherme Melo Esteves [1], João Esteves [1], Marta Resende [1], Luzia Mendes [1] and Andreia S. Azevedo [2,3,4,*]

1 FMDUP—Faculty of Dental Medicine, University of Porto, Rua Dr. Manuel Pereira da Silva, 4200-393 Porto, Portugal; guiesteves11@gmail.com (G.M.E.); up201404016@edu.fmd.up.pt (J.E.); martasantosresende@gmail.com (M.R.); lgoncalves@fmd.up.pt (L.M.)
2 LEPABE—Laboratory for Process Engineering, Environment, Biotechnology and Energy, Faculty of Engineering, University of Porto, Rua Dr. Roberto Frias, 4200-465 Porto, Portugal
3 i3S—Instituto de Investigação e Inovação em Saúde, Universidade do Porto, Rua Alfredo Allen 208, 4200-135 Porto, Portugal
4 IPATIMUP—Institute of Molecular Pathology and Immunology, University of Porto, Rua Júlio Amaral de Carvalho 45, 4200-135 Porto, Portugal
* Correspondence: asazevedo@fe.up.pt

Abstract: Regarded as one of the best solutions to replace missing teeth in the oral cavity, dental implants have been the focus of plenty of studies and research in the past few years. Antimicrobial coatings are a promising solution to control and prevent bacterial infections that compromise the success of dental implants. In the last few years, new materials that prevent biofilm adhesion to the surface of titanium implants have been reported, ranging from improved methods to already established coating surfaces. The purpose of this review is to present the developed antimicrobial and antibiofilm coatings that may have the potential to reduce bacterial infections and improve the success rate of titanium dental implants. All referred coating surfaces showed high antimicrobial properties with effectiveness in biofilm control, while maintaining implant biocompatibility. We expect that by combining the use of oligonucleotide probes as a covering material with novel peri-implant adjuvant therapies, we will be able to avoid the downsides of other covering materials (such as antibiotic resistance), prevent bacterial infections, and raise the success rate of dental implants. The existing knowledge on the optimal coating material for dental implants is limited, and further research is needed before more definitive conclusions can be drawn.

Keywords: oral biofilm; dental implants; titanium implants; antimicrobial; surface coating; anti-fouling

Citation: Esteves, G.M.; Esteves, J.; Resende, M.; Mendes, L.; Azevedo, A.S. Antimicrobial and Antibiofilm Coating of Dental Implants—Past and New Perspectives. *Antibiotics* **2022**, *11*, 235. https://doi.org/10.3390/antibiotics11020235

Academic Editor: Domenico Schillaci

Received: 8 January 2022
Accepted: 8 February 2022
Published: 11 February 2022

1. Introduction

Replacing missing teeth with dental implants is one of the most common treatment options with a great success rate. However, they still fail a significant number of times due to infections such as peri-implant mucositis, a biofilm-induced inflammation that can trigger bone loss, and result in peri-implantitis [1–8].

The success of oral rehabilitation using dental implants depends on numerous factors. The implantation process requires good interactions between the titanium surface and surrounding bone tissue (osseointegration), as well as resistance against bacterial colonization since implant-related infections are responsible for a large part of implant failure [9].

Studies indicate that around 29.48% (implant-based) and 46.83% (subject-based) of dental implants suffer from peri-implant mucositis and around 9.25% (implant-based) and 19.83% (subject-based) develop peri-implantitis [10]. Peri-implant mucositis is a biofilm-induced inflammation localized on the soft peri-implant mucosa, without any evidence of supporting bone loss (Figure 1B) [8]. It develops from healthy peri-implant mucosa around osseointegrated dental implants after the accumulation of bacterial biofilms. The major clinical sign of peri-implant mucositis is bleeding on probing (BOP), although it can also present erythema, swelling, and suppuration [2]. Clinical studies reported reversibility of peri-implant mucositis state after at least three weeks of better oral hygiene and biofilm control [3]. Nonetheless, if left untreated, the inflammatory process may progress and trigger the gradual destruction of the bone surrounding the implant, resulting in peri-implantitis [8].

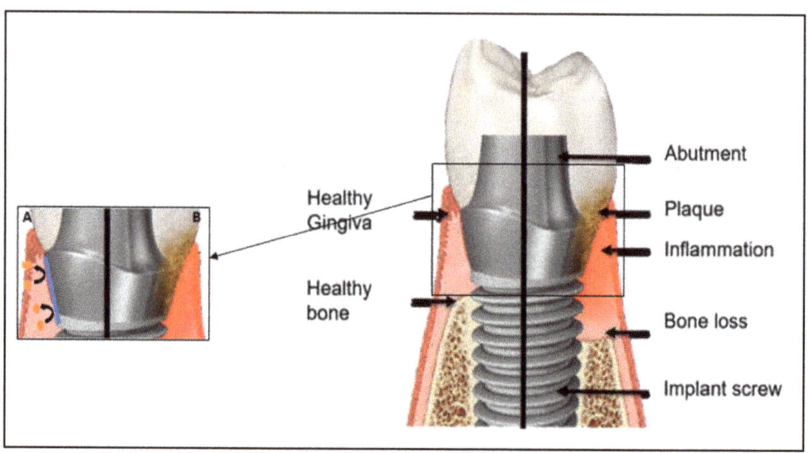

Figure 1. Biofilm formation process around dental implants and dental structures. (**A**) Implant coating prevents microbial colonization. (**B**) Dental plaque formation.

These types of infections, when untreated, result in implant loosening and require implant removal [4]. The ideal dental implant should have both great osseointegration properties and protection from the bacteria that cause peri-implant mucositis [2,3]. To achieve that, titanium and its alloys were chosen as the go-to material in commercialized implants due to its properties such as great resistance to corrosion, biocompatibility, and good tolerance by the biological environment, amongst others [2,3,5,7].

In order to improve success rates, modifications to the implant surface were proposed and studied. Changing the surface's physic and chemical characteristics such as roughness, surface free energy, and wettability allowed improving osseointegration [5]. The next step was to control and prevent the accumulation of bacteria around the implant since bacterial adhesion occurs immediately after implantation and results in biofilm formation. Biofilm is also resistant to many antimicrobial agents, making it difficult to treat once established [11]. In order to prevent bacterial infections, surface coatings with antimicrobial properties were hypothesized to be a reliable solution for this problem.

Titanium implants are susceptible to bacterial adhesion (Figure 1B) depending on the implant surface [12]. In order to prevent bacterial colonization, the titanium surface may be treated by adding materials or agents in the form of coatings (Figure 1A) [11].

Coating materials such as silver, copper, zinc, chlorhexidine, and some antibiotics presented to be a promising solution due to antimicrobial properties that would fight bacterial colonization [4,9]. Nonetheless, the methods required to modify and incorporate coatings in the implant surface are complex and expensive. Furthermore, while trying to achieve maximum antimicrobial properties, biocompatibility and osseointegration properties may be lost. The balance will always be key to determining the potential of a coating [13]. An ideal implant should have both osseointegration and antimicrobial properties [4,14]. There are numerous coatings developed throughout the years with promising results enhancing antimicrobial properties, achieved by either physical or chemical modification or even a combination of both, and we will discuss the most relevant ones.

This review aims to examine a wide range of coating materials and procedures in order to determine which novel solutions offer the best chances of producing a viable anti-fouling surface coating for dental implants.

2. Materials and Methods

2.1. Search Strategy and Information Sources

Four electronic sources of evidence were consulted in search of suitable articles that matched the aim of this review: MEDLINE (via PubMed), Scopus (Elsevier: Amsterdam, The Netherlands), Web of Science (Clarivate Analytics: Philadelphia, PA, United States), and SciELO (Scientific Electronic Library Online: Brazil), up to 28 May 2021.

Papers were searched using the following keywords: "Dental implant"; "Periodont*"; "Periodontal pockets"; "Microbes"; "Oral biofilm"; "Micro *"; "Oral micro *"; "Bacteria"; "Surface coating"; "Antimicrobial"; "Antibacterial"; "Titanium implants"; "Antibiotic loaded coating; "Biosurfactant"; "Chlorhexidine coating"; "Polymer coating"; "Implant infections"; " Anti-infective biomaterials"; "Antiadhesive surfaces"; "Nanostructured materials"; "Antibiofilm molecules"; "Antisense peptide nucleic acids"; "Drug delivery"; and "Antimicrobial photodynamic therapy". The keywords were combined with Boolean operators "AND" or "OR" with proximity operators ["" and ()] and with the truncation operator (*) used whenever appropriate.

The electronic database search was supplemented with a hand search across the references of all included papers.

2.2. Eligibility Criteria

The search was designed to be as broad as possible, with the goal of including all studies that alluded to coatings treatments on titanium surfaces to test their applicability on dental implants. All studies with less than ten years of publishing data were included. Restrictions were made to article type by excluding reviews, thesis, case reports, or letters.

3. Results

The most notable coatings, as well as their most important qualities and properties, are described in this section.

3.1. Bacteriostatic Materials

Various molecules show bacteriostatic properties, which means that they can repel bacteria from the surface of the implant without killing it [9].

Polymers such as polycations and biosurfactants have been studied and applied to titanium surfaces, being able to provide bacteriostatic properties to titanium surfaces. Moreover, recent studies were able to combine bactericidal and bacteriostatic materials granting both properties to titanium surfaces [9].

3.1.1. Polymer Coatings

Polyethylene glycol (PEG) is one of the most widely used polymers that provide antifouling properties to material surfaces, namely titanium surfaces [15,16]. It has excellent bacteriostatic qualities due to its hydrophilic and flexible chains. However, the very efficient antibacterial repelling properties also inhibit eukaryotic cell attachment (e.g., osteoblasts), thus compromising osseointegration. Therefore, they require the addition of cell adhesive sequences such as RGD (arginine-glycine-aspartate) peptides to preserve their biocompatibility [15].

Aiming to prevent the adhesion of bacteria to the surface of medical implants, polymer coatings with hydrophobic polycations such as N,N-dodecyl,methyl-PEI, as described by Schaer et al. [17], were studied and not only have shown a significant reduction in bacterial colonies of *S. aureus* when coated in titanium surfaces in vitro but also when applied in sheep models in vivo. Membrane proteins, teichoic acids (Gram-positive bacteria), and negatively charges phospholipids (Gram-negative bacteria) grant a negative surface charge to microbial cells. Polycations are attracted to the negativity present in microbial cells' surface, and based on their amphiphilic properties, they can disrupt their membrane and enable cell lysis, and this results cell death, adding bactericidal potential to the polymer coating.

Unfortunately, fabrication of the coating structure is a costly and challenging process, and there is a risk of polymer degradation with time, which could compromise the long-term stability and effect of the coating surface. Moreover, some polymer coatings are not yet available for use in titanium dental implants since the process of screwing the implants to the bone would compromise the structure of the coating, making it an unviable option [9]. Nonetheless, when a stable structure of polymers and cell adhesive sequences is achieved, both anti-biofouling and osseointegration results are expected [15].

3.1.2. Totarol

Totarol is a natural antibacterial agent that presents to be a promising solution towards the prevention of biofilm formation [14]. Clinically efficient against *Methicillin-resistant Staphylococcus aureus* and with demonstrated low cytotoxicity, Totarol was hypothesized and tested as a coating surface for titanium implants [14].

Xu et al. [14] analyzed the behavior of Totarol coated titanium disks with *Streptococcus gordonii* and human saliva. After 24 h, all bacteria were killed when compared to the control group, demonstrating the bactericidal effect of Totarol against *S. gordonii*. When tested for the long-term antibacterial effect, it was noted that the bactericidal effect was weakened after 12 days, but bacteriostatic mechanisms, namely anti-adhesion and anti-aggregation, were still inhabiting *S. gordonii* proliferation on the titanium surface, even after 24 days, while maintaining biocompatibility.

Improvements can be made to this surface coating, mainly to the long-lasting efficiency of antibacterial properties. Nonetheless, Totarol is another promising candidate to prevent peri-implantitis in the healing stage of the implantation process [14].

3.1.3. Biosurfactants

Biosurfactants are the most recent addition to the list of possible coatings with antibacterial properties for dental implants. Tambone et al. [18] conducted the only study using rhamnolipids on a titanium surface.

Rhamnolipids are a microbial surfactant mainly produced by *Pseudomonas aeruginosa*. They can preserve the biocompatibility of the titanium surface due to their low cytotoxicity and restrain the microbial adhesion process due to their amphiphilic structure. They can also modify permanently cell membranes, which could result in cell lysis [18].

The coated titanium disk with 4 mg/mL of rhamnolipid solution was tested against *Staphylococcus aureus* and *Staphylococcus epidermidis* for 72 h. After 24 h, *S. aureus* inhibition was higher than 90%, and *S. epidermidis* inhibition ranged from 62 to 78% depending on titanium surface morphology. After 72 h, the reduction in *S. aureus* was about 7% and 10.3% for *S. epidermidis*. No cytotoxicity was verified on any coated surface [18].

Rhamnolipids seemed to be another promising strategy for reducing both bacterial adhesion and biofilm reduction on titanium surfaces.

3.2. Bactericidal Materials

Some of the known strategies to lower bacterial load include damage to the bacteria's membrane or cell wall, penetration of the cell wall, DNA damage that hinders bacteria multiplication, creation of reactive oxygen species (ROS), blocking of ATP synthase, and stopping cell respiration [9,19]. Some materials imbued in surface coatings can grant bactericidal properties to titanium dental implants through some of the mechanisms mentioned and prevent biofilm formation.

3.2.1. Antimicrobial Peptides (AMP)

Antimicrobial peptides are a potential solution against biofilm colonization on titanium dental implants due to their antimicrobial properties. Geng et al. [20] studied engineered chimeric peptides with antimicrobial activity and concluded that, despite the need for further studies, these peptides had promising results regarding antimicrobial activity. Zhou et al. [21] studied a cationic antimicrobial peptide, GL13K, and by using

X-ray photoelectron spectroscopy and ultrasonication showed that this AMP improved both antibacterial and cytocompatibility properties of titanium implants, greatly inhibiting biofilm growth in vitro of *P. gingivalis* cultures, in the first 12 h, when compared to the control group. After 72 h, the antibacterial effect of GL13K coated surfaces was less effective but still an improvement when compared to uncoated titanium surfaces.

AMPs can be a good alternative to commonly used antibacterial materials such as silver, due to their flexibility, since they possess both antibacterial and osseointegration properties. Despite having a broad spectrum of action against bacteria, they appear to have a lower propensity to develop antibacterial resistance and toxicity [15,16].

Despite promising results, bioactive coatings with AMPs require complex designs of synthetic peptides that are quite costly to fabricate, which may compromise their broad use in titanium dental implants [9].

3.2.2. Ion-Implanted Surfaces

Ions from elements such as fluorine (F), copper (Cu), zinc (Zn), chlorine (Cl), iodine (I), selenium (Se), or cerium (Ce) can be incorporated into coatings in titanium implants [9]. Additionally, Bismuth (Bi) has recently been proposed as an antibacterial addition for calcium phosphate cement and titanium surfaces [22]. Zhou et al. [23] evaluated the potential of doped fluorine in Ti02/calcium-phosphate coatings (TiCP). With three different amounts of fluorine in the coating designated TiCP-F1 (least amount of fluorine), TiCP-F6, and TiCP-F9 (most amount of fluorine), they concluded that the TiCP-F1 coating had higher osteogenic properties than pristine (uncoated) titanium, but lacked antibacterial properties. On the other hand, TiCP-F6 and TiCP-F9 coatings had increased amounts of fluorine and showed significantly improved osteogenic and antibacterial properties.

One common coating applied to titanium surfaces is calcium-phosphate (CaP) due to its bioactive and osteoconductive properties [4]. Aranya et al. [4] modified CaP's surface by doping it with fluoride and zinc ions, both alone and combined. Fluoride is known for its bactericidal effect while Zinc is more associated with osseointegration promotion, despite also showing antibacterial properties [4].

They studied the effectiveness of this coating against *P. gingivalis* [24,25]. FZn-CaP coating had great results regarding inhibition of bacterial adhesion with ~88% reduction when compared to uncoated control disks in the first 72 h. F-Cap and Z-Cap coatings each had ~89% reduction in bacterial adhesion. After 7 days, biofilm reduction was significantly lower for both coatings. Zinc and Fluoride doped into CaP coating is a great option for dental implants since it enhances titanium surfaces with both bactericidal and bioactive properties [4].

Shen et al. [26] studied and verified that incorporating Zn ions in titanium dental implants surface coatings reduced the growth of *P. gingivalis*. Lin et al. [27] used Bismuth (Bi) to chemically modify titanium implants and was able to reduce *S. mutans* colonization.

A variety of tested ions also proved to be a promising solution to grant antimicrobial properties to the surface of titanium dental implants although they still lack long-term effects.

3.2.3. Photoactivatable Bioactive Titanium

Titania or titanium dioxide (TiO_2) is a nanocomposite coating with antimicrobial properties once it is photo-activated [9]. Under strong UV light, reactive oxygen species (ROS) are generated, which allows TiO_2 to kill a wide range of microorganisms such as bacteria while maintaining biocompatibility [9,28].

TiO_2 properties such as its' low-cost, stability, reactivity, durability, biocompatibility, and corrosion resistance make it a great option for commercial antimicrobial coatings. Thus, it is also possible to incorporate inorganic metals, such as copper or silver or even non-metals such as fluorine (F or Ca, particles previously mentioned), to enhance even further the antibacterial properties demonstrated by TiO_2 coatings [29].

3.2.4. Nanomaterials

Nanoparticles (NPs) are small particles with diameters between 1 and 100 nm from metals such as silver, gold, and other nanomaterials such as magnesium, zinc, or copper that display antimicrobial activity. Their antibacterial properties lead to research towards their incorporation in coatings for titanium implants [9,30].

The biocidal mechanisms shown by these nanoparticles, mainly the metallic ones, are diverse, which prevents bacteria in developing resistance against them [19]. Ag ions are released for a long period, expanding their antibacterial effect [31,32].

Massa et al. [33] incorporated Ag nanoparticles in a nanoporous silica coating through a sol–gel technique and observed a significant increase in both bactericidal and bacteriostatic properties of the titanium implant.

Silver nanoparticles (AgNPs) have shown a strong and wide antibacterial spectrum. Their exact mechanism against bacteria is still up for discussion, but the most accepted one so far is that AgNPs produce reactive oxygen species that inhibit the growth of bacteria, killing them in the process. For this reason, silveris one of the most used coating agents for titanium dental implants and other titanium medical devices [34]. However, some reports state that high concentrations of silver could be cytotoxic towards eukaryotic cells (e.g., fibroblasts and osteoblasts), which would reduce osseointegration properties of the implant [15]. Further studies are required to fully understand silver nanoparticles' behavior when coated in titanium implants.

3.2.5. Antibiotic Coatings

Nichol et al. [35] developed a single-layered sol–gel coating loaded with gentamicin on a titanium surface and tested it against *Staphylococcus* strains. Gentamicin is active against both Gram-negative and Gram-positive bacteria and is considered a broad-spectrum antibiotic.

Within 1 h, the Minimum Inhibition Concentration (MIC) was achieved, and after 24 h, all marked *Staphylococcus* variants were eliminated while 48 h later 99% of the gentamicin present in the coating was eluted [35].

These results were satisfactory to the author but do not represent the ideal coating for dental implants since antibiotic release was too fast for long-term prevention [35].

Zhang et al. [36] prepared titanium implants coated with vancomycin by using the electrospinning technique. Vancomycin was chosen due to its broad antimicrobial spectrum that covers both methicillin-resistant *S. epidermidis* as well as methicillin-resistant *S. aureus*.

The prepared coating showed an initial burst of vancomycin release on the first day (about 50.3%) followed by a slower and steadier release over the following 27 days (32.4%), making it a total release of approximately 528.2 μg of antibiotic from around 627.6 μg loaded in the coating (82.7%). No cytotoxicity to the cells was detected, and the antibacterial effect of Vancomycin was validated both in vitro and in vivo, showing promising results towards prevention of early implant-associated infections but still lacking long-term effects [36].

Lv et al. [37] also proposed an antibiotic-loaded coating to inhibit biofilm formation. They studied titanium substrates coated with a chitosan/alginate layer loaded with minocycline through layer-by-layer self-assembly. Minocycline is a broad-spectrum tetracycline antibiotic often used in conjunction with mechanical biofilm debridement in the treatment of periodontitis and peri-implantitis lesions.

This approach is extremely promising since the multilayered coating allows higher quantities of loaded antibiotics and a more controlled and over-time release of the substance.

The results obtained showed an initial burst of minocycline release in the first 24 h, which could fight the immediate colonization of bacteria. The antibiotic release stabilized during the first 7 days, and after that, the average concentration of minocycline on the fourteenth day was ~25.13 4.1 μg /mL. No bacterial cells with intact shape could be found on the titanium surface after 7 days [37].

In a recent systematic review by Souza et al. [38], all available references about antibiotic coated titanium surfaces were analyzed.

Out of those 33 articles, 11 used gentamicin, and 11 used vancomycin. Other antibiotics had three or fewer studies. *S.aureus* was the infection model of choice for 31 of the 33 studies [38].

Comparing the results obtained among all 33 articles, there was a big disparity from authors studying the same antibiotic. For example, in gentamicin-loaded coatings, bacterial reduction varied from ~5 up to ~99.9%. In vancomycin-loaded coatings, bacterial reduction ranged from ~45.3 up to ~99.2%. In three of the thirty-three studies, there were no reductions at all or even higher bacterial load in the tested group [38].

Bearing in mind the widespread and even contradictory range of results obtained and displayed in Table 1, as well as the scarce amount of data available, especially regarding human clinical data, there exists no consensual opinion regarding the best therapeutic approach for antibiotic-loaded coatings to prevent peri-implant infections [38]. There are also concerns towards toxicity and possible development of bacterial resistance, risks that should be avoided [15].

Table 1. Summary of results obtained with different antibiotics.

Antibiotic	Model	Efficiency	Reference
Gentamycin	*Staphylococcus* variants	~99% (24 h)	[36]
	S. aureus	From ~5 to ~99%	[39]
Vancomycin	*S. epidermidis and S. aureus*	Significant reduction (non-specified)	[37]
	S. aureus	From ~45.3 to ~99.2%	[39]
Minocycline	*S. aureus*	~99% (7 days) and ~80% (14 days)	[38]
	S. aureus	Non-reported	[39]

The fact that both gentamicin and vancomycin are not gold standards for the treatment of oral infection, since they act mostly on aerobic gram-negative bacilli, [38] result in the necessity to develop studies with antibiotics such as amoxicillin and metronidazole that would be more relevant for dental implants

Another important aspect to consider is that most of these studies were not conducted in the oral cavity or do not mimick its environmental conditions; thus, any conclusion regarding their behavior in dental titanium implants needs further studies [38].

3.2.6. Silane

Silane is commonly used to induce surface modifications through a process designated as silanization, which allows the covalent attachment of various molecules (peptides, polymers, or proteins, for example) to the titanium surface [9,39].

Despite being used mainly as an anchor, some silanes have shown biological activities themselves. Buxadera-Palomero et al. [39], in a recent review, studied silane triethoxysilylpropyl succinic anhydride (TESPSA), which presents both osteoinductive and antibacterial activity. These authors compared uncoated titanium disks and TESPSA-coated disks, by using the silanization process, in vitro, using Streptococcus sanguinis and Lactobacillus salivaris cultures and even dental plaque collected from one volunteer. They accessed both cytotoxicity and antibacterial activity, and the results obtained demonstrate no signs of cytotoxicity and a significant reduction in bacterial adhesion even after 4 weeks of incubation when compared to uncoated disks. However, the results showed differences between the mono-species models and oral plaque, which proves the importance of using more than one biofilm model in these studies.

Ultimately, TESPSA-coated titanium presented great potential for dental applications after presenting a great antibacterial effect for a prolonged period.

3.2.7. Nitride Coatings

Titanium nitride (TiN) is a material used to improve surface properties [9]. This material presents excellent chemical stability as well as resistance to corrosion and high temperatures while maintaining biocompatibility [9].

The antibacterial effect of a TiN and quaternized TiN (QTiN) coating surface on titanium was studied in vitro using *P. gingivalis* cultures [40]. The results obtained showed a significant reduction in bacterial coverage on TiN and QTiN coated surfaces after 4 h of culture. The uncoated group had 85.2% bacterial coverage while TiN-coated had only 24.22% and QTiN-coated only had 11.4% surface covered with bacteria after 4 h, while exhibiting good cell biocompatibility and promotion of osteoblast adhesion [40].

In another study, Ji et al. [41] found no antimicrobial effect in vitro by TiN against *P. gingivalis*; thus, further studies are required since the results are controversial.

3.2.8. Chlorhexidine Coatings

Chlorhexidine has been used together with mechanical debridement to improve the effectiveness of treatment against peri-implantitis [9].

Lauritano et al. [42] studied the effectiveness of a silicone coating containing chlorhexidine against microbes inside and outside the implant-abutment junction (IAJ). They achieved the coated surface by immersion of the abutment in the polysiloxane solution for 10 min followed by centrifugation and heat treatment.

After 24 h incubation following contact with a microbial pool of *S.aureus*, *Escherichia coli*, *Pseudomonas aeruginosa*, and *Candida albicans*, the results showed no living microbes in the internal part of coated implants [42].

Considering the different approaches of coating the inside of the implant, preventing microbial growth in the IAJ, chlorhexidine also had promising results against the agents responsible for peri-implant infections in the short term [42].

3.3. *New Perspectives on the Treatment of Peri-Implant Diseases*

3.3.1. Antisense Oligonucleotides (ASOs)

The antisense oligonucleotides (ASOs) are short fragments of a nucleic acid that can block a pre-defined target due to its complementarity. They were first described in 1978 by Zamecnik and Stephenson, who reported a blockage in viral replication and protein translation of a sarcoma virus RNA, after exposure to an antisense 13-nucleotide-long oligodeoxynucleotide in vitro [43]. Since this first generation of ASOs, several modifications have been made to overcome important limitations and enable clinical application such as incorporating 2′-O-methyl (2′-OMe) in the DNA backbone [44], connecting the DNA ribose ring by a methylene bridge between 2′-O and 4′-C atoms- locked nucleic acid (LNA) technology [45], or even using peptide nucleic acids (PNA) [46].

Antisense oligonucleotides can be used to interfere with essential biological processes of bacteria, which is helpful against bacterial infections. They can target many mRNA encoding essential genes, as well as functional domains of both 23S and 16S rRNA. Apart from targeting the essential mRNA and rRNA, ASOs can target non-essential genes related to biofilm formation, as many bacterial species form extracellular biofilms, making infections extremely challenging to eradicate. Some examples of biofilm-related genes are *motA* gene, encoding the element of the flagellar motor complex, the efaA gene, which plays an important role in the adhesion of bacteria to surfaces [46,47]. ASOs can also block two-component signal transduction systems, such as VicRK, that induce the gene expression for the synthesis of extracellular insoluble glucan, an extracellular matrix component of biofilms [48].

Biomaterial-based therapy has huge potential as biomaterial carriers can sustain slow-releasing drugs. Future development of new coating materials that can prevent the adhesion and subsequent biofilm formation of bacteria on dental implants, based on therapeutic antisense oligonucleotides, could be the key to fighting periodontitis. In 2020 Wu S. et al. developed a graphene oxide (GO)-based plasmid transformation system using electrostatic interacted GO-polyethyleneimine (PEI) complexes loaded with antisense vicR plasmid (GO-PEI-ASvicR). They showed that GO-PEI could efficiently deliver ASvicR plasmids into *S. mutans* cells with excellent transcripts of ASvicR, significantly reducing biofilm aggregation and exopolysaccharide (EPS) accumulation [48]. It is worth noting that graphene-based coating on titanium surfaces can be successfully obtained by electrodeposition as shown by Jankovic A (2015), making it a feasible option for innovative titanium coatings.

3.3.2. Bacteriophages (Phages)

Bacteriophages (phages) are bacteria-infecting viruses that can detect specific receptors in bacteria, inject their genetic material, and exploit the host's biochemical machinery to produce additional phage particles and enzymes, causing bacterial lysis. [49,50]. Shortly, the approach uses lytic pre-produced phages in a biodegradable drug delivery system to release and start their activity at the implant site [51]. Bacteriophages (phages) have emerged as a viable alternative to existing antimicrobial chemotherapy because of their ability to infect and kill specific bacterial strains while leaving the commensal microbiome intact [51]. The microbiota associated with a healthy (or commensal) state is more generalist, while disease-provoking microbiota is influenced by keystone microorganisms that have metabolic functions and an elevated virulence capacity that is mostly absent in healthy states [52]. As a result, we have reasons to believe that the use of bacteriophages (phages) would represent enhanced antimicrobial capacity, without compromising the microbiome associated with a healthy *periodontium*.

However, little evidence has been provided of the use of bacteriophages (phages) as therapy for dental implant-associated infections, limiting its application to the prevention and treatment of urinary catheters, respiratory ventilators, or orthopedic implant-associated infections [51,53,54].

3.3.3. Antimicrobial Photodynamic Therapy (aPDT)

The main purpose of peri-implant disease treatment is to disinfect implant surfaces as well as supporting tissues, and non-surgical and surgical mechanical debridements with ultrasonic scalers or periodontal curettes are regarded as essential techniques for this purpose [55]. However, none of these approaches have proven to remove or at least inactivate these peri-implant infections due to the macroscopic and largely microscopic intricacy of the implant's surface (rough and microporous) [56]. In addition to these methods, several studies have suggested that using adjuvant modalities, such as photodynamic therapy, can improve the treatment's outcome [57,58].

Some research has looked into whether a synergistic combination of aPDT and coating materials (such as chitosan) can work as a synergistic antimicrobial agent against bacteria that trigger peri-implantitis, such as *S. aureus*, *E. coli*, and *P. aeruginosa* [56].

Antimicrobial photodynamic therapy (aPDT) gained popularity in the early twentieth century as a result of the work of Herman von Tappeiner's team and is now used not only in medicine to treat certain tumors and skin diseases but also in dentistry to treat a variety of oral conditions such as peri-implantitis and peri-mucositis [58,59]. Photodynamic therapy (PDT) has been proved to be a successful treatment for peri-implantitis in the previous decade, owing to its ability to reach and penetrate the implant's uneven surface [56]. The treatment consists of a reaction between an innocuous, non-invasive, and non-toxic photosensitizer (such as methylene blue or toluidine blue) combined with a low-energy light source in the presence of oxygen. For them to react, the light must have a precise wavelength that corresponds to the photosensitizer's radiation absorption range, resulting in the creation of reactive oxygen species that are harmful to the bacterial cell and cause it to die. Gram-positive bacteria may be more vulnerable to this approach than Gram-negative bacteria due to the composition of their cell walls, making the photosensitizer more capable of invading those cells [60–62].

4. Discussion

4.1. Summary of Evidence

Our findings show a wide range of options and techniques to achieve an antimicrobial effect on the coating surface, as listed in Table 2. We cannot state which surface coating is the best based on the evidence since they all have their advantages and disadvantages.

However, we believe that progress is being made towards improved dental implant solutions, but the ideal one, which promotes cell adhesion, biocompatibility, and antibacterial action overtime at a fair cost, is still a few years away.

Table 2. Synthesis of the gathered evidence.

Coating Surface	Mechanism of Action	Major Upside(s)	Major Downside(s)
Polymer Coatings	Bacteriostatic (mainly)/Bactericidal	Great anti-biofouling and osseointegration properties when paired with cell-adhesive sequences; great bacteriostatic results in vitro	Risk of polymer degradation; require pairing with cell adhesive sequences
Antimicrobial Peptides	Bactericidal	Broad spectrum; low cytotoxicity; low propensity to develop antibiotic resistance	Complex structure; high cost of fabrication
Ion-implanted Surfaces	Bactericidal	Flexibility; can be paired with other coatings to promote both osseointegration and anti-biofouling properties	Difficulty to achieve a long-term antimicrobial effect
Photoactivatable Bioactive Titanium	Bactericidal	Cheap; stable; biocompatibility	Inability to photoactivate once the implantation occurs
Nanomaterials	Bacteriostatic (mainly)/Bactericidal	Longer antimicrobial effect	Efficiency is controversial; some studies report cytotoxicity
Totarol	Bacteriostatic (mainly)/Bactericidal	Efficient and long antimicrobial effect	Biodegradable substance
Antibiotic Coatings	Bactericidal	Cheap; good efficiency against targeted bacteria	Development of bacterial resistance; difficulty to achieve long-term release; toxicity
Chlorhexidine Coatings	Bactericidal	Great results in vitro regarding biofilm reduction	Absorption by the titanium surface
Biosurfactants	Bacteriostatic	Some bactericidal effects, increasing effectiveness	Scarce studies
Nitride Coatings	Bactericidal	Promotion of osteoblast adhesion while maintaining the antimicrobial effect	Controversial results against bacteria present in the oral cavity
Silane	Bactericidal	Combination of antibacterial effect and osteoinductive properties	Require further studies with different biofilm models
Antisense Oligonucleotides (ASOs)	Bacteriostatic	Can be used to interfere with essential biological processes of bacteria	The complex design of the probes to avoid low affinity to the target
Bacteriophages	Bactericidal	Having the ability to infect and kill specific bacterial strains while leaving the commensal microbiome intact	Little evidence has been provided in dental implant-associated infections

Antibiotic coating of dental implants is highly preferred over other options according to our research, mostly because this option is closely linked to dental implant placement, as antibiotics are frequently recommended as a prophylactic medication for this procedure [36]. Even though there are several diverse protocols with various antibiotics, dosages, and administration times, the existing literature supports the benefits of prophylactic antibiotic therapy against implant failure due to immediate bacterial colonization [63]. The success rate of implant placement is higher when a prophylactic antibiotic is administrated to the patient, however, it only affects the early colonization of bacteria to the implant surface, not preventing biofilm establishment in the following days, weeks, or years. Not to mention, the actual amount of antibiotic that reaches the site of the implant is lower when compared to local delivery of the same antibiotic [35,36].

A surface coating with a controlled antibiotic delivery system demonstrates to be a great long-term solution to control and prevent biofilm formation. Specific agents released over time that target early colonizers without compromising the mechanical, physical, and chemical properties of the dental implant and showing non-cytotoxic effects to host tissues and cells would drastically decrease the occurrence of peri-implant infections [38].

Antibiotic coatings, on the other hand, have some drawbacks, the most significant of which are related to antibiotic resistance. As a result of their antibacterial and antimicrobial capabilities, we believe that alternative coating materials should be preferred as a viable alternative against biofilm colonization on titanium dental implants.

More than 700 species of bacteria populate the oral ecosystem [9,34] with *Actinobacteria*, *Bacteroidetes*, *Firmicutes*, and *Proteobacteria* being the most relevant for oral health [9]. Regarding dental biofilm, which may vary between individuals and even among different sites of the oral cavity, a core microbiome was proposed, and it included the following species: *Streptococcus*, *Veillonella*, *Granulicatella*, *Rothia*, *Actinomyces*, *Prevotella*, *Capnocytophaga*, *Porphyromonas*, and *Fusobacterium* [34,52].

The process of biofilm formation starts with salivary glycoproteins developing a conditioning film (also known as the acquired enamel pellicle) on the teeth surface, which allows initial bacterial adherence [34]. After that process, weak long-distance forces between charged molecules of the pellicle and the pioneer bacterial species will grant initial adhesion. These forces grow stronger via receptor pairs between adhesins in the bacteria's surface and glycoprotein receptors in the acquired pellicle [34]. After the initial adhesion, biofilm development continues with cell aggregation until a stable microcolony is achieved. At last, due to multiple factors such as lack of nutrient or fluid dynamics, biofilm can disperse from the surface of the implant and migrate to other areas or tissues [64].

Medical devices, among others dental titanium implants, show a process of biofilm formation quite similar to natural teeth. A titanium surface, when present in the oral cavity, is immediately coated by plasma and saliva proteins. This will result in the formation of a protein layer that allows initial colonizers, such as various species of *Streptococcus*, to bind to it. Co-aggregation follows and interactions by different species induce biofilm accumulation. Finally, the extracellular matrix starts embedding microbial communities, and the biofilm is established [65].

Some materials such as silver have been used over the past few years as a coating material to reduce bacterial infections, but its use has been decreasing over time [15] due to concerns related to toxicity, which resulted in the switch to other solutions such as titanium dental implants with modified surfaces to improve osseointegration properties along with systemic administration of antibiotics, as we already discussed [15].

For the reasons and advantages stated, we propose that oligonucleotide probes should be considered as a feasible option for coating surface implants. We also suggest that future research should focus on determining whether the application of a combination of two or three different coating materials on the surface of dental implants can provide a synergic antimicrobial and antibacterial effect. Novel peri-implant therapies will result in a new and improved therapeutic approach and cannot be disregarded as an adjuvant approach of surface coatings. We anticipate that, by using these methods, we will be able to avoid the drawbacks of alternative coating materials (such as antibiotic resistance), prevent bacterial infections, and increase the success rate of dental implants.

4.2. Limitations

There is a lot of effort being placed into the discovery and applicability of established and new antimicrobial materials, but coating methods are complex and expensive.

Most of the mentioned coating possibilities are only applied in in vitro studies, and the ones that are employed in these conditions are not enough to establish clear conclusions. With so few in vivo trials, it is difficult to say when an efficient anti-fouling coating surface that would not drive up the price of a titanium dental implant will exist. We consider that more in vivo studies using relevant animal models and studies over longer periods are required for a better understanding of what is viable and what is not.

Many of the materials presented in this review were used as surface coatings for orthopedic implants. Despite being similar in most aspects, there are still quite a few relevant differences regarding orthopedic and dental implants. The oral cavity and oral microbiota are unique and distinct from the rest of the body, as we stated, and many of the results obtained require further investigation mimicking the environmental conditions of the oral cavity before they become possible solutions for dental implants.

5. Conclusions

All mentioned agents in this review have shown high levels of bacterial reduction when coated to titanium surfaces in vitro. However, the cost of fabrication, duration of the effect, and loss of osteoinductive properties appear to be the biggest obstacles faced to their broad appliance in titanium implants. Another major concern is the paucity of information on the bioactive surfaces' long-term durability after implantation. The majority of the described methods have excellent outcomes, but only for the first 24 to 48 h following implantation. Some approaches, such as antibiotic-loaded coatings applied layer-by-layer, address this issue, but they are insufficient and require additional development to be a viable option.

Antibiotic coatings have shown the most promising results so far when taking into consideration the duration and antimicrobial effect combined with anti-fouling and antibiofilm properties. This coating material is the most popular; however, it faces a major drawback associated with the emergence of antibiotic resistance.

In conclusion, the evidence about ideal dental implants' coating material is scarce and further studies are required before presenting more consolidated conclusions. More high-quality randomized clinical trials (RCTs) with longer follow-up periods, more precise criteria, and better-described coating protocols oriented to oral-biofilm-induced diseases are needed to establish a standardized guideline for this therapy's application. It is also critical to conduct more research comparing the application of titanium dental implant's coatings adjuvant to other complementary treatments for peri-implant diseases, such as antimicrobial photodynamic therapy, in order to establish which ones offer the most benefits. Furthermore, we highlight that antimicrobial and antibiofilm coatings applied to the surface of dental implants must not harm the microbiota associated with a healthy *periodontium*.

Author Contributions: Conceptualization, G.M.E., L.M. and J.E.; methodology, G.M.E., L.M., J.E., M.R. and A.S.A.; data curation, L.M., J.E., M.R. and A.S.A.; writing—original draft preparation, J.E. and G.M.E.; writing—review and editing, L.M., J.E., M.R., G.M.E. and A.S.A.; supervision, L.M. and M.R.; funding acquisition, A.S.A. All authors have read and agreed to the published version of the manuscript.

Funding: This work was financially supported by the following: Base Funding—UIDB/00511/2020 of the Laboratory for Process Engineering, Environment, Biotechnology and Energy—LEPABE—funded by national funds through the FCT/MCTES (PIDDAC); and Project POCI-01-0145-FEDER-030431, funded by FEDER funds through COMPETE2020—Programa Operacional Competitividade e Internacionalização (POCI)—and by national funds (PIDDAC) through FCT/MCTES.

Conflicts of Interest: The authors declare no conflict of interest.

References

1. Daubert, D.M.; Weinstein, B.F. Biofilm as a risk factor in implant treatment. *Periodontol. 2000* **2019**, *81*, 29–40. [CrossRef] [PubMed]
2. Heitz-Mayfield, L.J.A.; Salvi, G.E. Peri-implant mucositis. *J. Periodontol.* **2018**, *89*, S257–S266. [CrossRef] [PubMed]
3. Salvi, G.E.; Aglietta, M.; Eick, S.; Sculean, A.; Lang, N.P.; Ramseier, C.A. Reversibility of experimental peri-implant mucositis compared with experimental gingivitis in humans. *Clin. Oral Implants Res.* **2012**, *23*, 182–190. [CrossRef] [PubMed]
4. Kulkarni Aranya, A.; Pushalkar, S.; Zhao, M.; LeGeros, R.Z.; Zhang, Y.; Saxena, D. Antibacterial and bioactive coatings on titanium implant surfaces. *J. Biomed. Mater. Res. A* **2017**, *105*, 2218–2227. [CrossRef]
5. Santiago-Medina, P.; Sundaram, P.A.; Diffoot-Carlo, N. Titanium Oxide: A Bioactive Factor in Osteoblast Differentiation. *Int. J. Dent.* **2015**, *2015*, 357653. [CrossRef]
6. Derks, J.; Tomasi, C. Peri-implant health and disease. A systematic review of current epidemiology. *J. Clin. Periodontol.* **2015**, *42*, S158–S171. [CrossRef]
7. Bermejo, P.; Sanchez, M.C.; Llama-Palacios, A.; Figuero, E.; Herrera, D.; Sanz Alonso, M. Biofilm formation on dental implants with different surface micro-topography: An in vitro study. *Clin. Oral Implants Res.* **2019**, *30*, 725–734. [CrossRef]
8. Xi, D.; Wong, L. Titanium and implantology: A review in dentistry. *J. Biol. Regul. Homeost. Agents* **2021**, *35*, 63–72.
9. Chouirfa, H.; Bouloussa, H.; Migonney, V.; Falentin-Daudre, C. Review of titanium surface modification techniques and coatings for antibacterial applications. *Acta Biomater.* **2019**, *83*, 37–54. [CrossRef]
10. Lee, C.T.; Huang, Y.W.; Zhu, L.; Weltman, R. Prevalences of peri-implantitis and peri-implant mucositis: Systematic review and meta-analysis. *J. Dent.* **2017**, *62*, 1–12. [CrossRef]

11. Jemat, A.; Ghazali, M.J.; Razali, M.; Otsuka, Y. Surface Modifications and Their Effects on Titanium Dental Implants. *Biomed. Res. Int.* **2015**, *2015*, 791725. [CrossRef] [PubMed]
12. Thukkaram, M.; Coryn, R.; Asadian, M.; Esbah Tabaei, P.S.; Rigole, P.; Rajendhran, N.; Nikiforov, A.; Sukumaran, J.; Coenye, T.; Van Der Voort, P.; et al. Fabrication of Microporous Coatings on Titanium Implants with Improved Mechanical, Antibacterial, and Cell-Interactive Properties. *ACS Appl. Mater. Interfaces* **2020**, *12*, 30155–30169. [CrossRef] [PubMed]
13. Janson, O.; Gururaj, S.; Pujari-Palmer, S.; Karlsson Ott, M.; Stromme, M.; Engqvist, H.; Welch, K. Titanium surface modification to enhance antibacterial and bioactive properties while retaining biocompatibility. *Mater. Sci. Eng. C Mater. Biol. Appl.* **2019**, *96*, 272–279. [CrossRef] [PubMed]
14. Xu, Z.; Krajewski, S.; Weindl, T.; Loeffler, R.; Li, P.; Han, X.; Geis-Gerstorfer, J.; Wendel, H.P.; Scheideler, L.; Rupp, F. Application of totarol as natural antibacterial coating on dental implants for prevention of peri-implantitis. *Mater. Sci. Eng. C Mater. Biol. Appl.* **2020**, *110*, 110701. [CrossRef]
15. Mas-Moruno, C.; Su, B.; Dalby, M.J. Multifunctional Coatings and Nanotopographies: Toward Cell Instructive and Antibacterial Implants. *Adv. Healthc. Mater.* **2019**, *8*, e1801103. [CrossRef]
16. Liu, Z.; Liu, X.; Ramakrishna, S. Surface engineering of biomaterials in orthopedic and dental implants: Strategies to improve osteointegration, bacteriostatic and bactericidal activities. *Biotechnol. J.* **2021**, *16*, e2000116. [CrossRef]
17. Schaer, T.P.; Stewart, S.; Hsu, B.B.; Klibanov, A.M. Hydrophobic polycationic coatings that inhibit biofilms and support bone healing during infection. *Biomaterials* **2012**, *33*, 1245–1254. [CrossRef]
18. Tambone, E.; Bonomi, E.; Ghensi, P.; Maniglio, D.; Ceresa, C.; Agostinacchio, F.; Caciagli, P.; Nollo, G.; Piccoli, F.; Caola, I.; et al. Rhamnolipid coating reduces microbial biofilm formation on titanium implants: An in vitro study. *BMC Oral Health* **2021**, *21*, 49. [CrossRef]
19. Fanoro, O.T.; Oluwafemi, O.S. Bactericidal Antibacterial Mechanism of Plant Synthesized Silver, Gold and Bimetallic Nanoparticles. *Pharmaceutics* **2020**, *12*, 1044. [CrossRef]
20. Geng, H.; Yuan, Y.; Adayi, A.; Zhang, X.; Song, X.; Gong, L.; Zhang, X.; Gao, P. Engineered chimeric peptides with antimicrobial and titanium-binding functions to inhibit biofilm formation on Ti implants. *Mater. Sci. Eng. C Mater. Biol. Appl.* **2018**, *82*, 141–154. [CrossRef]
21. Zhou, L.; Lai, Y.; Huang, W.; Huang, S.; Xu, Z.; Chen, J.; Wu, D. Biofunctionalization of microgroove titanium surfaces with an antimicrobial peptide to enhance their bactericidal activity and cytocompatibility. *Colloids Surf. B Biointerfaces* **2015**, *128*, 552–560. [CrossRef] [PubMed]
22. Chen, F.; Liu, C.; Mao, Y. Bismuth-doped injectable calcium phosphate cement with improved radiopacity and potent antimicrobial activity for root canal filling. *Acta Biomater.* **2010**, *6*, 3199–3207. [CrossRef] [PubMed]
23. Zhou, J.; Li, B.; Han, Y. F-doped TiO_2 microporous coating on titanium with enhanced antibacterial and osteogenic activities. *Sci. Rep.* **2018**, *8*, 17858. [CrossRef] [PubMed]
24. Lafaurie, G.I.; Sabogal, M.A.; Castillo, D.M.; Rincon, M.V.; Gomez, L.A.; Lesmes, Y.A.; Chambrone, L. Microbiome and Microbial Biofilm Profiles of Peri-Implantitis: A Systematic Review. *J Periodontol.* **2017**, *88*, 1066–1089. [CrossRef]
25. Socransky, S.S.; Haffajee, A.D.; Cugini, M.A.; Smith, C.; Kent, R.L., Jr. Microbial complexes in subgingival plaque. *J. Clin. Periodontol.* **1998**, *25*, 134–144. [CrossRef]
26. Shen, X.; Hu, Y.; Xu, G.; Chen, W.; Xu, K.; Ran, Q.; Ma, P.; Zhang, Y.; Li, J.; Cai, K. Regulation of the biological functions of osteoblasts and bone formation by Zn-incorporated coating on microrough titanium. *ACS Appl. Mater. Interfaces* **2014**, *6*, 16426–16440. [CrossRef]
27. Lin, D.J.; Tsai, M.T.; Shieh, T.M.; Huang, H.L.; Hsu, J.T.; Ko, Y.C.; Fuh, L.J. In vitro antibacterial activity and cytocompatibility of bismuth doped micro-arc oxidized titanium. *J. Biomater. Appl.* **2013**, *27*, 553–563. [CrossRef]
28. Kiran, A.S.K.; Kumar, T.S.S.; Sanghavi, R.; Doble, M.; Ramakrishna, S. Antibacterial and Bioactive Surface Modifications of Titanium Implants by PCL/TiO_2 Nanocomposite Coatings. *Nanomaterials* **2018**, *8*, 860. [CrossRef]
29. Kumaravel, V.; Nair, K.M.; Mathew, S.; Bartlett, J.; Kennedy, J.E.; Manning, H.G.; Whelan, B.J.; Leyland, N.S.; Pillai, S.C. Antimicrobial TiO2 nanocomposite coatings for surfaces, dental and orthopaedic implants. *Chem. Eng. J.* **2021**, *416*, 129071. [CrossRef]
30. Parnia, F.; Yazdani, J.; Javaherzadeh, V.; Maleki Dizaj, S. Overview of Nanoparticle Coating of Dental Implants for Enhanced Osseointegration and Antimicrobial Purposes. *J. Pharm. Pharm. Sci.* **2017**, *20*, 148–160. [CrossRef]
31. Choi, S.H.; Jang, Y.S.; Jang, J.H.; Bae, T.S.; Lee, S.J.; Lee, M.H. Enhanced antibacterial activity of titanium by surface modification with polydopamine and silver for dental implant application. *J. Appl. Biomater. Funct. Mater.* **2019**, *17*, 2280800019847067. [CrossRef] [PubMed]
32. Wu, S.; Li, A.; Zhao, X.; Zhang, C.; Yu, B.; Zhao, N.; Xu, F.J. Silica-Coated Gold-Silver Nanocages as Photothermal Antibacterial Agents for Combined Anti-Infective Therapy. *ACS Appl. Mater. Interfaces* **2019**, *11*, 17177–17183. [CrossRef] [PubMed]
33. Massa, M.A.; Covarrubias, C.; Bittner, M.; Fuentevilla, I.A.; Capetillo, P.; Von Marttens, A.; Carvajal, J.C. Synthesis of new antibacterial composite coating for titanium based on highly ordered nanoporous silica and silver nanoparticles. *Mater. Sci. Eng. C Mater. Biol. Appl.* **2014**, *45*, 146–153. [CrossRef] [PubMed]
34. Liu, P.; Hao, Y.; Zhao, Y.; Yuan, Z.; Ding, Y.; Cai, K. Surface modification of titanium substrates for enhanced osteogenetic and antibacterial properties. *Colloids Surf. B Biointerfaces* **2017**, *160*, 110–116. [CrossRef] [PubMed]

35. Nichol, T.; Callaghan, J.; Townsend, R.; Stockley, I.; Hatton, P.V.; Le Maitre, C.; Smith, T.J.; Akid, R. The antimicrobial activity and biocompatibility of a controlled gentamicin-releasing single-layer sol-gel coating on hydroxyapatite-coated titanium. *Bone Jt. J.* **2021**, *103-B*, 522–529. [CrossRef] [PubMed]
36. Zhang, L.; Yan, J.; Yin, Z.; Tang, C.; Guo, Y.; Li, D.; Wei, B.; Xu, Y.; Gu, Q.; Wang, L. Electrospun vancomycin-loaded coating on titanium implants for the prevention of implant-associated infections. *Int. J. Nanomed.* **2014**, *9*, 3027–3036.
37. Lv, H.; Chen, Z.; Yang, X.; Cen, L.; Zhang, X.; Gao, P. Layer-by-layer self-assembly of minocycline-loaded chitosan/alginate multilayer on titanium substrates to inhibit biofilm formation. *J. Dent.* **2014**, *42*, 1464–1472. [CrossRef]
38. Souza, J.G.S.; Bertolini, M.M.; Costa, R.C.; Nagay, B.E.; Dongari-Bagtzoglou, A.; Barao, V.A.R. Targeting implant-associated infections: Titanium surface loaded with antimicrobial. *iScience* **2021**, *24*, 102008. [CrossRef]
39. Buxadera-Palomero, J.; Godoy-Gallardo, M.; Molmeneu, M.; Punset, M.; Gil, F.J. Antibacterial Properties of Triethoxysilylpropyl Succinic Anhydride Silane (TESPSA) on Titanium Dental Implants. *Polymers* **2020**, *12*, 773. [CrossRef]
40. Camargo, S.E.A.; Roy, T.; Carey Iv, P.H.; Fares, C.; Ren, F.; Clark, A.E.; Esquivel-Upshaw, J.F. Novel Coatings to Minimize Bacterial Adhesion and Promote Osteoblast Activity for Titanium Implants. *J. Funct. Biomater.* **2020**, *11*, 42. [CrossRef]
41. Ji, M.K.; Park, S.W.; Lee, K.; Kang, I.C.; Yun, K.D.; Kim, H.S.; Lim, H.P. Evaluation of antibacterial activity and osteoblast-like cell viability of TiN, ZrN and (Ti1-xZrx) N coating on titanium. *J. Adv. Prosthodont.* **2015**, *7*, 166–171. [CrossRef] [PubMed]
42. Lauritano, D.; Bignozzi, C.A.; Pazzi, D.; Cura, F.; Carinci, F. Efficacy of a new coating of implant-abutment connections in reducing bacterial loading: An in vitro study. *Oral Implantol.* **2017**, *10*, 1–10. [CrossRef] [PubMed]
43. Zamecnik, P.C.; Stephenson, M.L. Inhibition of Rous sarcoma virus replication and cell transformation by a specific oligodeoxynucleotide. *Proc. Natl. Acad. Sci. USA* **1978**, *75*, 280–284. [CrossRef] [PubMed]
44. Manoharan, M. 2′-Carbohydrate modifications in antisense oligonucleotide therapy: Importance of conformation, configuration and conjugation. *Biochim. Biophys. Acta* **1999**, *1489*, 117–130. [CrossRef]
45. Vester, B.; Wengel, J. LNA (locked nucleic acid): High-affinity targeting of complementary RNA and DNA. *Biochemistry* **2004**, *43*, 13233–13241. [CrossRef]
46. Wojciechowska, M.; Równicki, M.; Mieczkowski, A.; Miszkiewicz, J.; Trylska, J. Antibacterial peptide nucleic acids—Facts and perspectives. *Molecules* **2020**, *25*, 559. [CrossRef]
47. Baker, B.F.; Monia, B.P. Novel mechanisms for antisense-mediated regulation of gene expression. *Biochim. Biophys. Acta* **1999**, *1489*, 3–18. [CrossRef]
48. Wu, S.; Liu, Y.; Zhang, H.; Lei, L. Nano-graphene oxide with antisense vicR RNA reduced exopolysaccharide synthesis and biofilm aggregation for Streptococcus mutans. *Dent. Mater. J.* **2020**, *39*, 2019–2039. [CrossRef]
49. Dedrick, R.M.; Guerrero-Bustamante, C.A.; Garlena, R.A.; Russell, D.A.; Ford, K.; Harris, K.; Gilmour, K.C.; Soothill, J.; Jacobs-Sera, D.; Schooley, R.T. Engineered bacteriophages for treatment of a patient with a disseminated drug-resistant Mycobacterium abscessus. *Nat. Med.* **2019**, *25*, 730–733. [CrossRef]
50. Kingwell, K. Bacteriophage therapies re-enter clinical trials. *Nat. Rev. Drug Discov.* **2015**, *14*, 515. [CrossRef]
51. Kaur, S.; Harjai, K.; Chhibber, S. Bacteriophage mediated killing of *Staphylococcus aureus* in vitro on orthopaedic K wires in presence of linezolid prevents implant colonization. *PLoS ONE* **2014**, *9*, e90411. [CrossRef] [PubMed]
52. Esteves, G.M.; Pereira, J.A.; Azevedo, N.F.; Azevedo, A.S.; Mendes, L. Friends with benefits: An inside look of periodontal microbes' interactions using fluorescence in situ hybridization—Scoping review. *Microorganisms* **2021**, *9*, 1504. [CrossRef] [PubMed]
53. Barros, J.A.R.; de Melo, L.D.R.; da Silva, R.A.R.; Ferraz, M.P.; de Rodrigues Azeredo, J.C.V.; de Carvalho Pinheiro, V.M.; Colaço, B.J.A.; Fernandes, M.H.R.; de Sousa Gomes, P.; Monteiro, F.J. Encapsulated bacteriophages in alginate-nanohydroxyapatite hydrogel as a novel delivery system to prevent orthopedic implant-associated infections. *Nanomed. Nanotechnol. Biol. Med.* **2020**, *24*, 102145. [CrossRef] [PubMed]
54. Prazak, J.; Valente, L.; Iten, M.; Grandgirard, D.; Leib, S.L.; Jakob, S.M.; Haenggi, M.; Que, Y.-A.; Cameron, D.R. Nebulized bacteriophages for prophylaxis of experimental ventilator-associated pneumonia due to methicillin-resistant *Staphylococcus aureus*. *Crit. Care Med.* **2020**, *48*, 1042–1046. [CrossRef] [PubMed]
55. Romanos, G.E.; Javed, F.; Delgado-Ruiz, R.A.; Calvo-Guirado, J.L. Peri-implant diseases: A review of treatment interventions. *Dent. Clin.* **2015**, *59*, 157–178.
56. Camacho-Alonso, F.; Salinas, J.; Sánchez-Siles, M.; Pato-Mourelo, J.; Cotrina-Veizaga, B.D.; Ortega, N. Synergistic antimicrobial effect of photodynamic therapy and chitosan on the titanium-adherent biofilms of *Staphylococcus aureus*, *Escherichia coli*, and *Pseudomonas aeruginosa*: An in vitro study. *J. Periodontol.* **2021**. [CrossRef]
57. Novaes, A.B.; Ramos, U.D.; de Sousa Rabelo, M.; Figueredo, G.B. New strategies and developments for peri-implant disease. *Braz. Oral Res.* **2019**, *33*, e071. [CrossRef]
58. Mang, T.; Rogers, S.; Keinan, D.; Honma, K.; Baier, R. Antimicrobial photodynamic therapy (aPDT) induction of biofilm matrix architectural and bioadhesive modifications. *Photodiagn. Photodyn. Ther.* **2016**, *13*, 22–28. [CrossRef]
59. Chambrone, L.; Wang, H.L.; Romanos, G.E. Antimicrobial photodynamic therapy for the treatment of periodontitis and peri-implantitis: An American Academy of Periodontology best evidence review. *J. Periodontol.* **2018**, *89*, 783–803.
60. Albaker, A.M.; ArRejaie, A.S.; Alrabiah, M.; Abduljabbar, T. Effect of photodynamic and laser therapy in the treatment of peri-implant mucositis: A systematic review. *Photodiagn. Photodyn. Ther.* **2018**, *21*, 147–152. [CrossRef]

61. Sivaramakrishnan, G.; Sridharan, K. Photodynamic therapy for the treatment of peri-implant diseases: A network meta-analysis of randomized controlled trials. *Photodiagn. Photodyn. Ther.* **2018**, *21*, 1–9. [CrossRef] [PubMed]
62. Tavares, L.J.; Pavarina, A.C.; Vergani, C.E.; de Avila, E.D. The impact of antimicrobial photodynamic therapy on peri-implant disease: What mechanisms are involved in this novel treatment? *Photodiagn. Photodyn. Ther.* **2017**, *17*, 236–244. [CrossRef] [PubMed]
63. Romandini, M.; De Tullio, I.; Congedi, F.; Kalemaj, Z.; D'Ambrosio, M.; Lafori, A.; Quaranta, C.; Buti, J.; Perfetti, G. Antibiotic prophylaxis at dental implant placement: Which is the best protocol? A systematic review and network meta-analysis. *J. Clin. Periodontol.* **2019**, *46*, 382–395. [CrossRef] [PubMed]
64. Jia, G.; Zhi, A.; Lai, P.F.H.; Wang, G.; Xia, Y.; Xiong, Z.; Zhang, H.; Che, N.; Ai, L. The oral microbiota-a mechanistic role for systemic diseases. *Br. Dent. J.* **2018**, *224*, 447–455. [CrossRef] [PubMed]
65. Mohanty, R.; Asopa, S.J.; Joseph, M.D.; Singh, B.; Rajguru, J.P.; Saidath, K.; Sharma, U. Red complex: Polymicrobial conglomerate in oral flora: A review. *J. Fam. Med. Prim. Care* **2019**, *8*, 3480–3486.

MDPI
St. Alban-Anlage 66
4052 Basel
Switzerland
Tel. +41 61 683 77 34
Fax +41 61 302 89 18
www.mdpi.com

Antibiotics Editorial Office
E-mail: antibiotics@mdpi.com
www.mdpi.com/journal/antibiotics

www.ingramcontent.com/pod-product-compliance
Lightning Source LLC
LaVergne TN
LVHW070839170726
843515LV00005B/522

* 9 7 8 3 0 3 6 5 4 8 0 2 9 *